Rise of the Arab Tech Billionaires

Dive into the Blueprints of 25+ Most Profitable Business Models of the Arab Tech Startups

Dr. Uzma Farheen

Disclaimer

The information contained in this book is for educational and informational purposes only. It should not be interpreted as or substituted for professional business, financial, legal, or other advice. The trends, data, and perspectives shared in this work aim to provide insights into the Arab world, but they may not be applicable or relevant to every situation.

The author and publisher accept no responsibility for any direct, indirect, incidental, consequential, or punitive damages arising from the use of this material. Any reliance on the information contained within is strictly at the reader's own risk.

The fact that an individual, organization of website is referred to in this work as a citation and/or potential source of further information does not mean that the author endorses the information the individual, organization to website may provide or recommendations they/it may make. Further, readers should be aware that Internet websites listed in this work might have changed or disappeared between when this work was written and when it is read.

Dedication

This book is dedicated to my dear husband, my love and support, and my dear boy—Nikko, my Shih Tzu, my bundle of joy, who showed me the transformative power of love; my dear Elsa, a golden retriever who truly has a golden heart; and to all those who are struggling but keep moving despite hardships. Keep dreaming; if you can dream it, you can achieve it. Believe that there is a plan for everyone!

Preface

When I moved from India to Dubai, I was immediately struck by the rapid digital transformation happening across the Arab world. As a business consultant poring over research reports, I couldn't help but notice the intense focus by major firms on the technological awakening sweeping the Middle East and North Africa (MENA) region.

Technology is truly the new oil driving economic growth and innovation in this part of the world. From skyrocketing tech startup unicorns to widespread adoption of digital payment systems, the MENA nations are charting a bold new digital course. Algorithm-driven smart traffic management, ecommerce, fintech, and other advances are modernizing infrastructure and business at a startling pace.

In the following pages, I'll explore the cutting-edge companies, entrepreneurs, and disruptive business models powering this seismic digital shift. The Arab world is shedding antiquated business practices and wholeheartedly embracing the digital future. My hope is that this book provides inspiration and concrete examples for aspiring entrepreneurs to launch their own tech ventures across the region.

More than just economic progress, this digital transformation represents an unparalleled opportunity to uplift societies, improve lives, and create a more sustainable, prosperous future for all people across the MENA nations. May these powerful new technologies be leveraged as engines for positive change, human betterment, and global good.

The Arab world has arisen as a potent epicenter of 21st century innovation. Let this exploration open your eyes to the vast potential awaiting bold new strategies and solutions catalyzed by the internet, mobile, cloud, AI and other revolutionary digital forces.

Let the transformation begin.

Table of Contents

Part I - Digitizing the Future: Paving the Way for Transformation of the Arab World.

The world belongs to those who can imagine it, design it, and execute it.

– *HH Sheikh Mohammed Bin Rashid Al Maktoum*

CHAPTER 1

Dawn of a New Era

In 2019, a rocket scientist from MIT named Dr Georges Aoude founded an AI startup called Derq in Dubai. The company delivers innovative road safety solutions that help eliminate crashes and save lives.

In 2023, Derq partnered with Dubai Integrated Economic Zones (DIEZ) to implement fourteen AI Smart Pedestrian Crossing Systems across Dubai Silicon Oasis.

Derq has been recognized as an industry leader by the WEF and has received several awards, including AI Company of the Year at SXSW 2019 and Top Road Safety Innovator for Vision Zero in 2020 by Together for Safer Roads.

Khaled Zaatarah, a twenty-eight-year-old Palestinian raised in Saudi Arabia and the UAE, was sitting in the back of his blue kayak and drifting to sea near the Palm Jumeirah. The back of the kayak is a more comfortable spot than the front. The front makes rowing a bit more arduous. But as the kayak drifted, he switched his position to the front. He couldn't believe what he felt. He had more control over the kayak. He was no more going with the flow. That's when he realized he needed to take control of his life. He quit his high-paying MNC job and started 360VUZ.

360VUZ is not just a platform for sharing 360-degree videos; it's a window into Khaled's world. His team creates most of the videos on the app, capturing the energy of concerts, the thrill of sports events, and the luxury of high-end houses.

In 2022, the 23rd Virtual International Selection Panel selected Khaled as the Endeavor Entrepreneur.

Careem, Fawry, Swvl, Yalla, and Dubizzle are tech start-ups based in the Middle East that have quickly achieved the status of Unicorns, valued at more than a billion dollars.

From financial citadels like Dubai to ancient seats of power in Cairo, seismic shifts driven by science and computing technology transform business, economics, and society at breakneck speed.

This is an unprecedented inflection point fuelled by converging factors - young, tech-savvy populations hungry to solve local challenges, governments racing to build digital economies, unprecedented levels of investment capital flooding the region seeking the next unicorn, and, most critically, ambitious, innovative thinkers leveraging technology to uplift their communities with ingenuity and drive.

Many countries in the Arab world have developed comprehensive national digital transformation plans to guide their efforts.

These plans mainly include tactics and approaches to using technology to its fullest to advance and upgrade infrastructure, improve government services, encourage and enhance computer literacy skills, and cultivate innovation among various groups of people.

To encourage the economy to move toward multiple sources of income, the Kingdom of Saudi Arabia has come up with **Saudi Vision 2030** - a list of ambitious goals that strongly focuses on successful economic activities emerging through digital technology, improving online government services, encouraging innovation and the starting of businesses, and establishing the digital skills of Saudi Arabia's citizens.

An initiative by the UAE called '**We the UAE 2031**' highlights a path for a continuously developing UAE. This path will take ten years, focusing on social, economic, investment, and development aspects. The plan seeks to enhance the position of the UAE as a global partner and an attractive and influential economic hub, as well as showcase the UAE's triumphant economic model and the possibilities it offers to all international partners. Its predecessor, **UAE Vision 2021**, was formed to focus and reach new heights in different sectors, namely healthcare, education, sustainability, and infrastructure. It also included the UAE National Agenda initiative, which focused on the abovementioned sectors, emphasizing digital transformation to improve citizens' efficiency, competitiveness, and quality of life.

Bahrain Economic Vision 2030 initiative emphasizes increasing the economy's production of goods and services and economic diversification through digital transformation. This plan contains points on developing a digital society and economy, improving government services via the Internet, investing in physical and software-based digital infrastructure, and developing skills.

Qatar National Vision 2030 aims to turn Qatar into a knowledge-based economy motivated by innovation and human development. It also includes strategies for reinforcing digital infrastructure, promoting government services via the Internet, assisting in entrepreneurship and innovation in sectors such as technology and ICT, and improving the population's digital skills.

To achieve inclusive growth, 'The *Egypt Vision 2030*' plan highlights the country's objectives of developing economically and progressing socially by focusing on using technology to its fullest and modernizing infrastructure. Broadening the population's access to digital services, encouraging e-commerce and digital entrepreneurship, growing the youths' digital skills, and drawing investors into the ICT sector are a few of the plan's aims.

An initiative called '*Jordan Vision 2025*' contains the long-term objectives to help the country develop sustainably, creating an economy based on knowledge and fostering innovation as its primary focus. It includes the country's intent on building up digital infrastructure, broadening access to wide-bandwidth data transmission, advancing and improving services provided by the government via the Internet, and investing in ICT innovation and entrepreneurship to assist in the digital economy's growth.

The Digital Tunisia 2025 initiative will implement all-inclusive strategies to accelerate the digital transformation of Tunisia's economy and society. The plan aims to modernize ICT infrastructure, improve the population's education and digital skills, actively encourage e-government services, assist in digital entrepreneurship and innovation, and attract more investors to the ICT sector.

A strategic plan designed by the government called '*The Digital Morocco 2025*' emphasizes using digital technologies to their advantage to encourage social inclusion and economic development. The plan contains approaches to broadening the availability of services offered digitally, improving ICT infrastructure and connectivity, and promoting the development of skills and digital literacy. Entrepreneurship and innovation will also assist in the expansion of the digital economy.

The Sultanate's *Oman Vision 2024* initiative aims to improve the Omani population's digital skills, assist in entrepreneurship and digital innovation, popularize e-government services, and modernize ICT infrastructure. The plan also includes long-term goals for Oman to develop sustainably and attain economic diversification.

A *New Kuwait Vision 2035* plan enclosed a set of guidelines to turn Kuwait into a central regional financial and commercial hub. The plan emphasizes Kuwaiti citizens flourishing in a digital economy, building up digital infrastructure, improving government services via the Internet, pushing digital innovation and entrepreneurship forward, and enhancing their skills.

Management consultancies' forecasts predict that digital service marketplaces across the Middle East and North Africa (MENA) region will expand by over 20% annually until 2025, on pace to exceed over $100 billion as millions of new tech-literate Arab users come online annually via smartphones. E-commerce alone is projected to grow by 250%, reaching $28.5 billion in the same period.

International competition is spurring governments to incentivize such rapid adoption from citizens and businesses, with the Gulf countries leading policy reforms and investments. Bloomberg analysis suggests venture funding into MENA startups has tripled since 2016, nearing $2 billion in 2021. Over a third of the Arab population is already online, leveraging digital services daily across verticals.

Youth Driving Momentum

Arab millennials and Gen Z will comprise over 65% of the population by 2025. These 220 million digitally hyper-connected youth are enthusiastically adopting new technologies at twice the pace of global peers, providing the momentum lifting entire economies' digital value chains. They are also increasingly the creators rather than just consumers of the technologies proliferating business and society.

Visionary leaders across the region have explicitly focused education and immigration policies around technology fields, seeding favorable conditions for youth-led innovation to blossom. The historic bias towards public sector careers is also now reversing as ambitious graduates aspire to launch businesses or work for corporates leading regional digital transformation.

Hotspots Igniting Startup Explosions

Epicenters across the Arab world are driving momentum through targeted initiatives, policy incentives, and funding schemes catalyzing local startup ecosystems to address industry gaps. Government-backed accelerators, cloud computing zones, research grants, and incentive-laden Free Trade Zones catering to tech firms provide fertile testing grounds for agile ventures aiming for swift global impact.

The UAE leads adoption with centers like Dubai Internet City hosting titans like Microsoft, IBM, and Huawei, converging with legions of regional startups and entrepreneurs. Meanwhile, Saudi Arabia vows to have over 450,000 tech professionals working across next-gen cities like Neom, which promises to serve as a regional Silicon Valley. Egypt and Jordan have crafted reputations as skilled coder talent hubs feeding regional innovation for global clients.

Many countries in the Arab world are actively preparing and implementing plans to maximize the possibilities brought about by technology and increase economic growth, innovation, and society's development.

In making public administration more modern, enhancing the delivery of services, and encouraging the government to be more open, understandable, and accountable to the public, the key players are the digital approaches that the Arab world governments are taking. As governments invest more in digital transformation, citizens can anticipate improvements to a greater extent in their governments' organization, accessibility, and response times across the region.

To reduce the amount of paperwork, lessen bureaucratic hassles, and make the overall user experience better for citizens, Arab governments are turning several services, including permit and license applications, tax and bill payments, accessing healthcare and education services, and getting official documents into digital services.

To easily access government services and information via the internet, these e-government platforms are being created and used to function as centralized portals, giving citizens what they need to connect with various agencies and departments of the government. They allow them to access different services, submit their applications, track their requests, and obtain news regarding their transaction status. For citizens to and actively participate with the government, these platforms were made to be secure, easy to navigate, and able to be accessed from any place as long as the user has an internet connection.

Electronic Document Management Systems (EDMS) are used to turn government documents and records digital and to manage them electronically. Digitally storing, organizing, and recovering documents and information are the primary uses of EDMS, lessening the risks of losing or damaging documents and reducing the need for paper-based filing systems. They also allow collaboration and information sharing among government departments, enhancing decision-making processes and efficiency.

The primary purpose of digital government approaches is to improve government operations' transparency and accountability by giving citizens more access to information and data. Policies that allow the public better access to government data, or open data policies, are being implemented, and platforms are being created to give the public a more transparent and accessible way of obtaining this information. Citizens monitor government activities through these policies, and they can track the expenses of the public and ensure that government officials are held responsible for how they act.

To make citizens engage and participate more in policy-making, digital government approaches use social media, online forums, and mobile apps to their advantage. These digital channels are used to gain feedback and comments on policy issues and involve citizens in decision-making. This way, the governments are sure that the policies and services being given meet the needs and priorities of the people they serve.

To sum up, a significant shift in the workings of different industries and sectors is happening to put tech at scale into action to create value consistently—digital Transformation.

We try to work only with the dreamers. This place is not for conventional people or companies.

– *Saudi Crown Prince Mohammed bin Salman*

CHAPTER 2

Tech Evolution in the Arab World

The Middle Eastern region has recently experienced a rise in tech businesses, which have promptly become the springboards for enhancing innovation, entrepreneurship, and technology. Whether in booming urban areas or budding localities, these tech hotspots are propelling digitalization in the region and positioning the Arab world as a force to be reckoned with on the tech stage. Let's take a cursory look at the principal locations of interest.

Dubai

Dubai has an advantageous location. At the heart of the UAE, it also boasts a favorable business climate and progressive leadership.

Dubai is the Middle East's foremost innovation and technology hub. There are many tech start-ups, incubators, and accelerators in Dubai, not to mention the numerous tech giants that have made the city the location of their regional headquarters.

Dubai's government has executed many schemes to act as a backbone for tech entrepreneurship. The Dubai Future Foundation and Dubai Internet City (DIC) are such programs offering financial support, physical space, and mentorship for tech industry entrepreneurs. DIC is a trailblazer in technological advancement in the heart of the United Arab Emirates. DIC was created in 2000 and has since transformed Dubai into a global center for technological innovation.

The DIC is the region's principal business hub and offers a beautiful atmosphere for tech hotspots to work together and grow. With its cutting-edge infrastructure, top-tier facilities, and thriving community of start-ups, international enterprises, and founders, DIC offers a favorable business environment for tech-driven entrepreneurs.

One principal attraction of DIC is its prime location within Dubai's bustling commercial area. It is located along Dubai Media City and Dubai Knowledge Park, both prominent free trade zones. DIC's location in such an area makes it a beneficiary of a network of industries that fosters collaboration and brainstorming.

Many tech firms in various sectors, including software development, telecommunications, artificial intelligence, blockchain, and e-commerce, are based in DIC. From established multinationals to brand-new start-ups, DIC attracts companies worldwide seeking to take advantage of Dubai's superb location, business-friendly atmosphere, and easy access to international markets. It promotes internal and external cooperation as companies come together to address complicated issues and inspire innovation. Through tech-specific ideas and partnerships, DIC boosts knowledge-sharing, technological transfer, and joint research and development.

In addition to the above, DIC also proffers broad auxiliary services to help companies excel. These services are assistance with start-up facilitation, networking seminars, mentorship/training programs for founders, access to inception funding, and investment opportunities.

Apart from its numerous draws, DIC drives engagement and talent improvement through diverse workshops, hackathons, and training. Its goal is to nurture the next generation of tech talent and build a culture of ingenuity.

Riyadh

The Saudi Arabian government introduced the *Vision 2030 initiative*, a scheme crafted to diversify the economy and enhance innovation. It is no new fact that, due to this scheme, Riyadh is now a noteworthy tech hub in the Middle East. There are many start-up incubators, research institutions, innovation centers, and an ever-increasing community of tech founders and investors.

Riyadh's welcoming business start-up environment is bolstered by State funding, legal reforms, and enterprises like the Future Investment Initiative, which has attracted global investment and talent to Saudi Arabia. Thus, spurred on by progressive State programs and a culture of entrepreneurship, Riyadh has seen unprecedented development.

Saudi Arabia's Vision 2030, an all-encompassing scheme to tackle sole reliance on oil and diversify the economy, has tremendously impacted Riyadh's tech growth. The vision's core focus is entrepreneurship, innovation, and the growth of a knowledge-based economy.

Saudi Arabia's government has executed various schemes to support entrepreneurship and ingenuity in Riyadh. These schemes include, but are not limited to, the National Transformation Program, the Future Investment Initiative, and the Public Investment Fund. All three provide finance, mentorship, and support services to tech start-ups.

Recently, Riyadh has witnessed a rise in venture capital investment as local and international investors have shown interest in the region's tech ecosystem. Venture capital firms like Riyadh Valley Company, Wa'ed Ventures, and STV have essential roles in funding novel tech businesses and inducing technological development. The State has also assigned funds to boost the growth of SMEs through schemes like the Saudi Venture Capital Company, the Saudi Industrial Development Fund, and the Small and Medium Enterprises General Authority.

Numerous tech hubs, co-working spaces, and incubators in Riyadh provide tech founders access to resources, mentorship, and networking opportunities. Some examples include Riyadh Valley Company, King Abdullah University of Science and Technology Entrepreneurship Centre, and the Badir Programme for Technology Incubators.

Riyadh presently boasts of an educated and skilled workforce with a rising number of STEM graduates. It also boasts several leading educational research institutions, such as King Saud University, Princess Nourah bint Abdulrahman University and King Fahd University of Petroleum and Minerals. These academic institutions help in developing the talent pool for the start-up ecosystem.

Saudi Arabia's government has also set up various initiatives that promote entrepreneurship and ingenuity via diverse means, which include entrepreneurship courses, training programs, and hackathons. Corporate collaborations are increasing in Riyadh's tech ecosystem as established organizations seek to evolve and stay competitive in this digital age. Corporate accelerators, innovation labs, and open innovation initiatives encourage partnerships between new businesses and established companies, thus driving the adoption of technology and market expansion.

Furthermore, the government has introduced regulatory legal reforms to boost entrepreneurship and Innovation in Riyadh. These reforms include legal measures to simplify business registration, streamline licensing procedures, and offer attraction incentives for tech start-up businesses.

Schemes like the Saudi Authority for Intellectual Property and the General Authority for Small and Medium-scale Enterprises are created to induce a more conducive legal environment for start-ups.

In addition, Saudi Arabia has claimed the top regional spot for electronic government service maturity for the second consecutive year in a United Nations ranking. The UN Economic and Social Commission for Western Asia's 2023 Government Electronic and Mobile Services Maturity Index recently analyzed 17 countries across 84 priority government services available through portals and applications.

Saudi Arabia scored high points across all the assessment criteria, reflecting astronomical development in service availability, user satisfaction, and public outreach in contrast with 2022.

With a growing number of new tech businesses, investors, and support institutions, Riyadh is well on its way to becoming a prominent hub for entrepreneurship and ingenuity in the Middle East and possibly globally.

Cairo

The Egyptian government set new plans to boost entrepreneurship in the area of tech, and that is why Egypt's capital city, Cairo, has now emerged as one of the foremost tech cities in the Middle East region.

With the city's enormous population, rising interest of investors, and active start-up atmosphere, the government introduced several funding programs, innovation hotspots like the Greek Campus, and novel legal reforms. As a result, Cairo now boasts of a thriving tech community with collaborating partnerships and more than enough tech talent from leading Egyptian universities and research organizations.

MAGNiTT, a start-up data platform, reported that the start-up sector in Egypt saw an exponential growth of about 176%, thus reaching a high record venture capital funding of $491 million. Still, according to the report, Egypt contributed to over 15% of transactions and 11% of deployed capital in the Middle Eastern and North African regions, with e-commerce and financial technology emerging as the most active fields. A popular data research organization, Start-Up Genome, has also placed Egypt as one of the world's top 10 emerging start-up areas, thus attributing a total value of $2.7 billion to its growing entrepreneurship landscape.

Beirut

The city of Beirut has battled several socio-political and financial problems over the years. However, it has become one of the dynamic mainstream tech locations in the Middle East. Beirut is known for its accommodating start-up environment and thriving entrepreneurial conditions. There are several tech start-ups, accelerators, joint working spaces, and an ever-increasing community of founders and investors.

The government of Beirut is not to be left out, as it has launched several programs to support tech initiatives. Such programs include Berytech and Beirut Digital District, providing seed finance, mentorship, and networking opportunities for tech start-ups. Berytech plays a multidimensional role as it constructs an all-inclusive atmosphere for the local founders while contributing its quota to developing the Lebanese entrepreneurial environment. The Berytech program has helped shape the societal norms to imbibe and assist the entrepreneurial mindset while pushing novel concepts. The Berytech programs also help connect diverse tech stakeholders and influence "tech-benefitting" policies and legislation.

Amman

Located in Jordan, this city is gaining wide acclaim as one of the promising tech hotspots in the Middle East. This is probably due to its skilled workers, supportive entrepreneurial legal regime, and improved tech environment. Amman also has a plethora of innovation centers, jointly used workspaces and accelerators. There are also several founders and investors. The Jordan Government has started implementing several regulatory programs and financial initiatives and setting up tech parks like the King Hussein Business Park to support tech ingenuity.

Doha

The Government of Qatar released its Qatar Vision 2030 Initiative, which has bolstered Doha's emergence to prominence as a location that promotes technology and its subsequent emergence as a popular tech hotspot as well as acting as a boost to the diversification of the economy. Doha has a strategic location, and with support from the Government, primarily via programs like the Qatar Science and Technology Park

(QSTP) and Qatar Development Bank (QDB), the city has become a top destination for tech talents, investors, incubators, and accelerators.

Tunis

Not many places in North Africa have emerged as major tech hotspots like Tunis. With its trained workforce, booming environment, and governmental support, the city hosts several joint workspaces, incubators, and start-ups. Tunis also accommodates a rising number of tech investors and talents. Of course, the State has helped set up programs that can boost tech ingenuity and entrepreneurship, especially with innovation hubs like Cogite.

Casablanca

Casablanca is another city in Northern Africa that has gained acclaim as a raving spot for start-ups and innovation. Because it boasts a great location, a growing economy, and entrepreneurial support from the state, it is perfect for tech growth, especially in fintech, agritech, and e-commerce.

These are all supported by the Technopark and Moroccan Agency for Digital Development (ADD) and other state initiatives. The Government in Casablanca invests in digital infrastructure and education to nurture the next generation of tech talent and promote the city as a top destination for tech innovation in the region.

Kuwait City

Kuwait City is fast emerging as a tech entrepreneurship and investment hub in the Gulf region. With its significant population, proper business regulatory regime, and Tech-driven interest, Kuwait City hosts a range of start-ups, tech incubators, and accelerators. In Kuwait City, the

Kuwait National Fund for Small and Medium Enterprise Development and the Kuwait Investment Authority (KIA) are in force as programs to support digital infrastructure, tech education, and collaborations between founders and investors.

The emerging tech hotspots in these cities show the dynamic environment in the Arab world that is furthering tech innovation, entrepreneurship, and financial growth across the region. These locations have supportive government policies, access to finance, and a rising tech talent pool. Each hub is adequately positioned to become a significant player in the international tech sphere.

I am delighted to celebrate the notable innovations of Omani youth across various sectors.

– *HH Sayyid Theyazin*

CHAPTER 3

The Booming Startup Scene

The emergence of unicorns in the Arab region is a testament to the robust start-up environment there. This is why the world's eyes are on the Middle East, as global investors have taken notice and have positioned the Middle East and North Africa (MENA) region as a hub of innovation and economic development.

A research report by a "Big Three" consulting firm has foreseen that almost 350 start-ups in the MENA region can achieve near-unicorn status by 2030 with a $1 billion valuation. The area is dedicated to building a booming start-up environment, making it stand out. If this consulting firm's forecast should materialize, it would connote the presence of disruptive, high-value companies across diverse industries.

Factors That Can Improve Start-up Success

There are a plethora of factors that can indicate success for a start-up. The factors listed below are founded on a comprehensive evaluation of start-ups all over the globe, both successful and unsuccessful. The factors identified are:

1. ***Market Need and Fit***: Successful start-ups usually identify a societal problem or need in the market and then create a product that fills that gap succinctly. To correctly identify opportunities in the market and create a product that fits, it is expedient to understand customer pain points, preferences, and behaviors.

2. ***Strong Value Proposition***: For a start-up to succeed, it must proffer a strong value proposition that differentiates its product from competitors and resonates with its target customers. A strong value proposition shows the target customers the unique benefits of the start-up's offer and generates demand in the market.

3. ***Customer-Centric Approach***: The product the start-up is offering must be centered on the customer's comfort and satisfaction. This is why successful start-ups must listen to customer needs, build their product based on customer likes, and prioritize customer feedback. Strong customer relationships and good customer service are essential drivers of long-term success for start-ups.

4. ***Scalability***: Tech start-ups need to be able to scale. If the organization cannot scale, it will most likely not succeed, as only scalable start-ups can grow and make a profit over time. Most successful start-ups design scalable business models that allow

them to proliferate and capture market shares without incurring cost increases.

5. ***Strong Leadership and Team:*** Successful start-ups are usually led by visionary founders and entrepreneurial leaders with clear vision, strong leadership skills, and the ability to inspire their teams. It is essential to have a talented team with complementary skills and expertise to execute the start-up's strategy and promote growth.

6. ***Agility and Adaptability:*** Most start-ups operate in dynamic and uncertain atmospheres. As a result, they need to be flexible enough to adapt to changes in the market and competition from other start-ups. Successful start-ups must experiment, learn from their failures, and quickly refine their offers and products.

7. ***Access to Resources and Networks:*** Human resources and capital are essential for start-up success. A start-up that will grow to become a force to contend with in the business world should be able to harness talent and access capital. These are needed to inspire growth, attract top talent to join their teams and leverage networks to access customers, suppliers, and investors.

8. ***Execution and Focus:*** Apart from planning, it is essential to follow up with execution. Every successful start-up needs to start with clear goals and execute its plans with discipline and focus. A start-up must minimize distractions, not be too fixated on past successes, and remain determined until it reaches the top.

9. ***Innovation and Differentiation***: Successful start-ups differentiate themselves by introducing new products as solutions to old market needs and new solutions to new problems. Innovation can take different forms, including technological breakthroughs, procedure improvements, or new approaches to addressing customers' needs. Start-ups that prioritize innovation and continuously strive to stay ahead of the curve usually remain at a competitive advantage.

10. ***Financial Management and Sustainability:*** Any start-up that intends to be successful must manage its finances prudently, maintain tight control over its expenses, and allocate resources effectively to maximize returns. Building a sustainable business model that can consistently generate revenue streams, efficiently manage cash flow, and remain profitable over time is necessary.

11. ***Resilience and Persistence:*** Start-ups usually face many challenges on their journey to success. These challenges include market volatility, competition, financial constraints, and internal disputes. To overcome these challenges, the founder and his executive team must navigate uncertainty, overcome hurdles, and persevere in adversity. They must learn from their failures, adapt to change, and stay committed to their long-term vision and goals.

12. ***Ethical and Responsible Practices:*** Today, several ethical rules and practices exist in different industries. To be a successful start-up, the team must play by the book and prioritize integrity, transparency, and social responsibility in product manufacturing and stakeholder interactions. Building trust with

customers, employees, investors, and the wider community is essential, as this will lead to a constant positive reputation and foster long-term relationships that will later contribute to sustainable business growth.

It is easy to be carried away by instant success or dedication to product success, forgetting customer satisfaction, ethical integrity, and general outlook. Incorporating the above principles will help start-ups navigate challenges, capitalize on opportunities, and build resilient, thriving businesses that create value for customers and society.

CHAPTER 4

The Anatomy of a Tech Start-Up

A nything that will have a lasting impact must be built on hard work, grit, and excellence. An intriguing story lies behind most tech successes, so much so that it has become a standard blueprint. There are usually different elements orchestrated to siphon value from innovative ideas and then move from ideation to commercial viability and, afterward, to sustained leadership. However, some principal steps are extremely important for every start-up that intends to be successful, as adherence to these steps can take such a business from being merely promising to a scalable and profitable operation capable of transforming industries. This will break down secret patterns behind successful tech businesses as we look into each component imbibed and followed by unicorns and emerging elites within the start-up lifecycle.

Phase 1 - Concept Origination

Every new success story can be traced back to simple insight, which later becomes an idea and addresses a problem or market need. Tech start-ups address this need via technology. This initial insight can be due to the following:

- **Inspiring Mentors**—When someone is exposed to bold thinkers, their mind is usually aroused to think entrepreneurially during his/her formative years, and this later blossoms into a great idea. John Hennessy, the famous professor, pioneer of computer architectural concepts, and future CEO of Alphabet, coached many students, including Instagram founder Kevin Systrom, at Stanford University.
- **Life Challenges**—We encounter several daily struggles that can ignite innovative solutions if we analyze them objectively. Jessica Chen Hua, the co-founder of Open Door, once lost her home rental, and that was when she realized that a gap needed to be filled in the area of property management. She then decided to solve the problem using FinTech tools.
- **Market Needs**—Researchers and market observers can spot problems, bottlenecks, and frictions that technology can solve in different industries. After identifying such market needs, the objective is to follow up by crafting proper tech solutions.

Phase 2 - Execution Planning

Once an idea is revealed, the concept must be translated into an actionable blueprint. To do this, the team needs to procure essential resources that will enable them to release a proper technological solution that will be the best fit. The following are the vital areas to think about here;

- **Finance**—Capital funds the manufacturing of a product and even the hiring of personnel. Conventional financial sources include personal savings, government grants, angel investments, bank loans, or venture capital. Every entrepreneur should note that the budget needs to align with long-term strategic financing.
- **Personnel**—It is essential to bring different individuals with specialized expertise and a shared vision before adding talent for launch roles in technology, design, marketing, and operations. Psychometric testing can significantly improve your decisions while setting up a team.
- **Product Development**—The minimum viable product (MVP), the actual technological solution, must be built with agile software to allow early testing to validate product-market fit assumptions made without all the initial overengineering.
- **Legal Bases**—Intellectual property protection like trademarks, copyrights, and patents must be considered, as well as incorporation regulations to operate and protect proprietary assets. Data privacy regulations must also be adhered to.
- **Culture**—The organization's founding principles, mission statement, and team values are all part of its culture. Remote work flexibility, dressing, the individual lifestyle of staff, and project-based organizational structures are all factors that shape organizational culture.

Phase 3 - Gaining Traction

How can you gain traction and attract customers to your product? The truth is that securing early customer wins is essential as it can drive growth cycles and bring in more capital and talent. Tactical leverage points to gain traction include:

- **Network Building**—As an entrepreneur, it's important to attend events, participate in seminars/webinars in the industry, and leverage social media, podcasts, and so on. These will bring you clients, get you in the door, and connect you with people who refer you to others.
- **Market Testing**—You can offer free trials for your product while leaving out the premium features for paid users. This method tempts conversion while offering limited-time discounts for initial user cohorts to test demand response to pricing and capture feedback.
- **Iteration Agility**—Rapid build-measure-learn sprint cycles help products adapt to actual user activity, which can be analyzed through robust analytics. Similarly, structured customer interviews disclose the customer's pain points.
- **Conversion Optimization**—You can insert tweaks to signup flows, microcopy in landing pages, and calls-to-action on your site or in the usage of your product. This lifts app store and website conversion rates, which are used to maximize the top-funnel appetite in the sales pipeline.
- **Viral Loops** – several referral schemes incentivize sharing and social media connectivity, amplifying reach and app integrations with platforms with extensive product usage and adoption.

Phase 4 - Growth Increment

Suppose you need to sustain the use of your product in the market and keep scaling new heights as a tech entrepreneur. In that case, you must devise a strategic advantage to place you ahead of your competitors. It has to be something your competitors will find difficult to replicate.

- **Network Effects**—Metcalfe's law quantifies platform value growth. Most business models harness niche use cases around communication, content sharing, and commerce, as this allows more participants, thus disincentivizing shifts.

- **High Retention**—Specific designs can increase customer lifetime value through higher engagement, retention, and customer satisfaction. You can use algorithms to know precisely when dopamine hits when using your product and harness it.

- **Ecosystem Synergy**—When you partner with others, you link a wide range of users. This act provides users seamless connectivity across external tools widely used in target niches because this supports and confers advantages to integrated service bundles over individual apps.

- **Data Monetization**—As digital activity explodes, intelligence from engagement patterns, online transactions, and behaviors creates new value-creation avenues, including providing analytics or segments to partners, advertisers, and research firms.

Phase 5 – Sustainable Innovation

It is important that as growth happens, your organization matures past pioneer status into an institution that requires continually renewing competitive advantages through innovation, upgrading core products, and launching new targeted solutions.

- **Talent Development** - keep securing talent globally regardless of the level of skill you already have. Organize training programs to level up internal teams and cultivate organizational learning culture as it fuels continuous innovation, which exceeds the competition.

- **New Platforms**—It is essential to keep reinventing your virtual platforms and expanding your sites as this helps to retain old users and capture new audiences. So, keep upping your content and payment facilitation game.

- **Forward Vision**—A startup can stay grounded in the future despite increasing bureaucracy if it has muscular R&D divisions rapidly prototyping experimental tech. Successful start-ups have to think ahead.

- **Acquisitions**—Strategic mergers and acquisitions play an essential role in providing access to fast-tracked expertise or technology that startups cannot quickly build alone. Post-merger integrations also focus on preserving entrepreneurial strength while improving the founders' core visions.

While the results are based on probability, the guidelines above have proven to be growth-boosters blueprint principles that are fertile soil for cultivating industry-leading enterprises from dorm rooms and garages across Silicon Valley and now globally. The following historic tech titan might be awaiting its awakening.

The Team

A tech start-up's success often hinges on its team's strength and cohesion. This is an example of a team structure for a thriving tech start-up:

1.   Founders

The founders consist of the CEO, CTO, CFO, and COO. Sometimes, one individual takes on the position of both CTO and COO, but it is advisable to separate all four positions because each has varying core duties.

The CEO (Chief Executive Officer) provides the company's strategic direction, vision, and leadership. He is the primary founder and is in charge of affairs. The CTO (Chief Technology Officer) manages the creation of the company's products and ensures technical strategy and implementation.
The CFO (Chief Financial Officer) oversees the company's financial aspects, such as raising funds, creating budgets, and planning finances. The COO (Chief Operating Officer) handles day-to-day operations and ensures that the company runs smoothly and efficiently.

2.   Product Development Team

This team occupies a core part of the start-up's operations, including product managers, software engineers, UI/UX designers, and quality assurance testers. The PMs (Product Managers) usually define the product vision (the why of the product), prioritize features, and coordinate with development teams. The software engineers develop the company's products and code and implement features based on product requirements. The UI/UX designers create intuitive and visually appealing user interfaces, ensuring a seamless user experience; at the same time, the quality assurance testers conduct rigorous testing to identify and fix bugs and ensure product reliability and quality.

3.   Sales and Marketing Team

This team plays a crucial role by analyzing the market (customers), predicting their needs, and determining the most effective way to present the product to attract customers. The Chief Marketing Officer creates marketing plans to advertise the start-up's products and company brand. The sales representatives identify the target market and bring in new customers. They also build relationships with clients and prospective clients.

Marketing specialists execute campaigns across different channels, including digital marketing, content marketing, and social media. Customer success managers handle customer inquiries and support. They also ensure customer satisfaction and foster long-term relationships with clients.

4.   Operations and Support Team

The team members here may not be directly involved in the start-up's main tasks. Still, they are crucial in handling secondary responsibilities, allowing other teams to focus on their primary tasks. The operational managers supervise logistics and administrative tasks such as supply chain management, inventory control, and vendor relations. The HR staff is responsible for recruiting, onboarding, training, and managing personnel. The customer service agents offer clients technical help and problem-solving aid, handle queries, and fix problems promptly.

5.   Data and Analytics Team

This team comprises data scientists, business analysts, and data engineers. The data scientists analyze data to gain insight and inform decision-making, utilizing techniques like machine learning and predictive analytics. The business analysts interpret data to identify trends, opportunities, and areas for improvement while providing actionable recommendations to the management. The engineers design and maintain data infrastructure and ensure data availability, reliability, and security.

6.   Executive Advisors and Board Members

This last group comprises experienced professionals who provide advice and mentorship, investors who provide funding and networking opportunities, and a board of directors who oversee the start-up's operations.

Please note that this is a general framework. Consequently, the team's particular responsibilities and organization might change based on the start-up's characteristics, field, and level of development.

CHAPTER 5

You Can Be A "Techpreneur," Too!

S everal unicorns today started as simple ideas in people's minds. Who's to say your idea now cannot grow into a billion-dollar industry tomorrow?

Here's the truth—you can be a "techpreneur," too! Look at the success stories of some start-ups and be inspired to make a move.

Facebook

The social media giant Facebook is a mixed tale of ambition, innovation, and the transformative power of connecting people in previously thought impossible ways. Facebook's journey started in the early 2000s in the dorms of Harvard University, when the internet was rapidly evolving, but virtual social connectivity had not yet reached its full potential.

In the fall of 2003, Mark Zuckerberg, then a sophomore at Harvard University, started working on a project that would later change social media. Zuckerberg initially created Facemash, a site that allowed its users to compare and rate the beauty of other students' photos. Despite its controversial nature and the latter's shutdown by Harvard University's administration, Facemash laid the groundwork for what would become Facebook later.

Still a young man driven by the internet's potential to unite people, Zuckerberg envisioned a platform connecting students across universities, not just Harvard. This led to the formation of a platform called 'The Facebook'. It was initially an exclusive network for Harvard students, and it was launched on February 4, 2004, from Zuckerberg's dorm room. The virtual platform allowed users to create profiles, upload photos, and connect with other students.

The site's simplicity, focus on real identities, and exclusivity contributed to its immediate popularity. Within 24 hours, over 1,000 Harvard students had signed up, and in a month, more than half of the undergraduate population had a profile. The demand for Facebook rose, leading to its expansion to other Ivy League schools. Eventually, all the universities in the United States and Canada had access to it.

Facebook was later changed from "The Facebook" to "Facebook," and by the end of 2004, it had more than one million active users. The statistics of its rise caught the attention of venture capitalists, which is why the company received its first significant investment the next year, exacerbating its expansion. Facebook was opened to students in high school, and in 2006, it became available to anyone with a valid email address. This marked the beginning of its transformation into a global social network.

From a college dorm project to a global platform, Facebook's rise shows the power of innovation and motivates everyone to buckle up and pursue their visions. Zuckerberg and his team faced many problems. These problems include court cases, data privacy concerns, and the daunting task of upscaling the platform to serve billions of users. Despite these impediments, Facebook has committed to making the world a connected village.

Today, Facebook is above a mere social networking site. It is a global society that connects people from every part of the world. It also enables them to share their stories, express their identities, and mobilize for social causes. Facebook is not just about creating a company but also achieving a vision that has radically changed how we interact, communicate, and connect. It's an important reminder that it is possible to make a global impact with determination, innovation, and a desire to connect people.

Uber

The creation of Uber is another story of innovation and determination. It shows just how much a simple idea can revolutionize an entire industry. Indeed, it is a tale that shows the transformative power of technology and entrepreneurship in solving real-world problems.

In 2008, two men, Travis Kalanick and Garrett Camp, attended LeWeb, a yearly tech event held in Paris, France. On a snowy evening, they were stranded and unable to get a cab. This led to a conversation between the two men about creating a timeshare limo service that could be ordered via an online app.

The idea was both about transportation convenience and using technology to make urban mobility more efficient and accessible. On their return to San Francisco, they swung into action to bring this idea to life. By the following year, the idea had become UberCab, a luxury car service that could be requested with a few taps on a smartphone.

The first Uber model was simply about connecting riders with luxury vehicle drivers for hire. What set UberCab apart was its focus on technology, convenience, and user experience as the app revealed the car's location in real time, provided an estimated arrival time, and handled all payment transactions electronically, removing the need for physical cash. The service quickly became popular in San Francisco and appealed to tech-savvy people who appreciated its convenience and efficiency.

The brand Uber also faced challenges. The company faced regulatory hurdles, pushback from traditional taxi-driving services, and doubts about its impact on traffic and urbanization. Uber persisted while continually adapting and innovating. The company increased its service offerings with UberX, allowing average individuals to become drivers using their vehicles. This reduced the costs for riders and democratized the service. This game-changer made Uber accessible to a broader audience and transformed it from a luxury service to a mainstream transport option.

Uber's story is not just a new way of getting from point A to point B but about changing movement in urban areas. The founders saw an opportunity to fill a gap in the existing transport system and pursued it with determination. Their vision for Uber was driven by a belief in technology to solve problems and better lives.

Today, Uber operates in hundreds of cities worldwide, and it is always expanding its services to include food delivery, freight, and even perhaps the future of urban air transport. The company's move from a simple idea resulting from frustration to a global brand shows the effect of innovation, the spirit of entrepreneurship, and the ability to solve everyday problems. Uber's story can inspire aspiring entrepreneurs to dream big and use technology to change the world around them.

TikTok

Many people talk about TikTok and usually use it to allude to innovation, cultural impact, and the rise of a platform that has radicalized social media and entertainment for the current generation. The use of TikTok shows the power of creativity, the importance of understanding user behavior, and the potential of technology to connect people across the globe in new and exciting ways.

TikTok's move started in 2016 when a Chinese tech company, ByteDance, released a short-form video app called Douyin for the Chinese market. Immediately, ByteDance realized the platform's global appeal, so they decided to expand and released TikTok in 2017. It was re-released on a larger scale to meet the needs of the global entertainment market. The app allowed its users to create short and engaging videos that often featured music, dance, comedy, and other creative content that could be easily shared with friends.

What makes TikTok stand out from other social media platforms is its highly personalized algorithm that curates content for its users, making it incredibly easy for anyone to discover videos that match their interests. This user-centered approach, plus the app's easy-to-use editing tools, is the game changer for most young people who, incidentally, are most of the app's users, as it has made them more creative in crafting content. TikTok's growth was explosive, and the platform became a cultural phenomenon, influencing music, dance, and fashion trends on a universal scale. It also provided a launch pad for unknown artists and creators to gain fame and recognition, democratizing content creation automatically.

TikTok also faced many challenges. First, the platform faced scrutiny over data privacy, the spread of misinformation, and the potential for misuse, especially by young people. TikTok has several measures to address these issues, implement stricter content moderation policies, and improve user safety.

One of the most inspiring parts of TikTok's story is how it has served as a source of joy, learning, and networking during challenging times such as COVID-19. During the Coronavirus pandemic in 2020, TikTok was a go-to app for many people as it was a space for people to connect, share experiences, and find solace in humor and creativity when the world seemed at its worst.

TikTok has changed the online entertainment space and created a new way for people to interact, express themselves, and build virtual communities. TikTok reminds us of the potential to bring people together via innovation and create a platform beyond geographical and cultural boundaries.

In the end, the inspiring story of TikTok is about more than just an app's success; it's about empowering individuals to share their voices with the world. It shows how creativity and technology can be combined to create a space where everyone has the opportunity to be seen and heard.

Stepping Stones

These stories show that problems and hurdles are stepping stones to success. Anyone with a clear vision, determination, and a willingness to adapt will always be at the forefront of revolutionizing industries and positively influencing lives.

The founders of these companies saw opportunities, whereas others saw obstacles and challenges. They dared to dream big and remained resolute despite opposition.

It is rudimental to reflect on the stories of these unicorns and let them inspire you to embrace your entrepreneurial spirit, pursue your passions tenaciously, and grab any opportunity to make a difference. The world is full of opportunities for those willing to challenge the status quo.

Part II - Sectors Riding the Digital Wave

"The desert tells a different story every time one ventures onto it."

- Robert Edison Fulton Jr.

CHAPTER 1

FinTech – Where Finance Meets Innovation

The MENA region is still going through a period where innovation is being birthed on a high side. This is due to many factors, including but not limited to favorable legal and regulatory regimes, higher adoption of digitalization, and a booming entrepreneurial environment. Many areas have emerged as FinTech hotspots all over the Arab world, especially in countries like the UAE, Saudi Arabia, Egypt, and even Bahrain. These areas attract domestic and global players who wish to capitalize on the growing demand for innovative financial services in the MENA region.

In the Middle East, FinTech solutions range from digital banking platforms to peer-to-peer lending sites, mobile payment options, insurance technology, regulatory technology, and blockchain applications. Advances in these areas are formed to cater to the evolving preferences of tech-savvy customers and address the monetary needs of businesses and institutions operating in the region.

The world of finance and technology is in a state of rapid transformation. The term "FinTech" has evolved to represent a call for disruption, innovation, and transformation. It's a meeting point where financial services and cutting-edge technology converge, promising to reshape the industry as we know it.

As the financial industry continues to welcome digitalization in different forms, FinTech stands at the fore of the dynamic evolution - reshaping traditional norms and defining the future of finance. It represents an atmosphere filled with possibilities, from mobile payment solutions and blockchain apps to robo-advisors and peer-to-peer lending platforms.

As the financial industry embraces digitalization, FinTech is leading the charge, reshaping traditional norms, and defining the future of finance. In this dynamic landscape, your role is crucial. By leveraging artificial intelligence, big data analytics, and seamless user experiences, fintech unicorns are revolutionizing how individuals and businesses manage their finances, access capital, and engage with the global economy.

In the broad area of FinTech, agility and adaptability are paramount, as both startups and established institutions are in constant competition to discover new opportunities and close gaps in the market. The combination of finance and technology has opened the way to unprecedented financial inclusion, increased access to banking services, investment opportunities, and global financial literacy.

The FinTech revolution is still unfolding, and it will shape the future of finance by going beyond borders and ushering in an era of financial empowerment and innovation. Whether it's rethinking the concept of currency, transforming payment systems, or making investment options more accessible, FinTech embodies progress and the persistent pursuit of a more interconnected, inclusive, and efficient financial environment.

Youth demographics and the soaring digital presence are key propellers driving the expansion of digital financial services in the Arab region, a vast and dynamic market eager to adopt digital solutions for financial transactions.

The Arab youths are known globally for their love of technology and digitalization; thus, they are usually early adopters of digital financial services. They are used to conducting various aspects of their lives online, from socializing and entertainment to shopping and banking. With smartphones now being used in the region for average use, young Arabs increasingly rely on their mobile devices to access information, communicate, and manage daily activities, including financial transactions.

This growing digitalization has made the Middle East a fertile ground for adopting tech financial services and e-payment platforms as Arab youths seek convenience, speed, and accessibility in their economic interactions, preferring digital channels over traditional banking methods. Young Arabs embrace a digital-first approach to managing their finances and making payments.

The Coronavirus pandemic also induced a shift towards digitalization in financial services, as lockdowns and social distancing measures prompted customers to rely on digital channels for their banking needs. This unprecedented crisis showed the world the importance of digital resilience and agility and increased the adoption of digital financial solutions among Arab youths.

FinTech start-ups and traditional financial institutions sprung up in response to this demand and ramped up their efforts to cater to the needs of young customers in the Arab world. They are rolling out innovative products and services tailored to the preferences and lifestyles of this digital-loving generation.

Such products include mobile banking apps with intuitive interfaces, contactless payment solutions, and reward programs designed to appeal to tech-savvy users. The combination of young demographies and increasing digitalization presents an opportunity for the expansion of digital financial services and e-payment platforms in Arab states.

By leveraging technology and possessing a deep understanding of the evolving needs and preferences of the youth segment, financial service providers can tap into this vast market and drive greater financial inclusion and empowerment across the region. Here are a few case studies within the FinTech sector in the Arab world. Let us explore their unique strategies, success factors, and the reasons behind their significant funding traction.

Case Study: *TABBY A BUY-NOW-PAY-LATER (BNPL)*
PLATFORM

Tabby: Transforming E-Commerce Payments in the Middle East

Tabby is the "buy now, pay later" (BNPL) platform that has revolutionized e-commerce payments in the MENA region. It was founded by Hosam Arab, a serial entrepreneur who aims to simplify online shopping and enhance the purchasing experience for consumers in 2019.

Since then, Tabby has become a game-changer in the Arab FinTech space. Traditional payments like cash on delivery (COD) have long dominated the e-commerce scene, posing a significant challenge for online retailers regarding cash flow, customer trust, and order fulfillment.

Hosam Arab desired a more convenient and flexible payment style, so he introduced a BNPL platform that would enable consumers to shop online and pay in installments without the need for interest fees or credit cards.

Tabby has a straightforward and transparent business model. The platform allows customers to split their purchases into interest-free installments over a lengthy time—the platform partners with many e-commerce business owners to integrate its payment solution into their checkout processes. As a result, customers can choose Tabby as their payment platform and select their preferred installment plan, with options from 2 months to 2 years.

Tabby's innovative risk assessment style has completely changed how business owners handle transactions, creating a seamless and efficient solution for businesses and everyday customers.

By leveraging advanced algorithms and real-time data analysis, Tabby can approve customers in a matter of seconds, thus streamlining the purchasing process and eliminating delays.

The merchant can ship the order immediately when a customer approves it because Tabby will handle the payment. Tabby will pay the merchant upon delivery, thus ensuring that businesses receive timely remuneration for their goods and services.

From the customer's perspective, Tabby's payment process is simple and convenient. Customers can settle their transactions using credit cards, debit cards, or other traditional payment options.

Tabby deducts the payment from the customer's choice immediately after the item reaches them, thus providing a hassle-free shopping experience.

Tabby's approach profits merchants by reducing the risk of payment challenges and improving cash flow.

It also creates a better shopping experience for customers. The platform helps businesses drive sales and build customer loyalty with quick approvals and flexible payment options while providing customers with a convenient and secure way to shop online.

Using the Tabby platform translates to a bucket full of benefits for both consumers and business owners:

1. **Convenience and Flexibility:** Tabby allows consumers to shop online and pay for their purchases in installments. This makes it easier to afford high-ticket items and spread payments over time.

2. **Interest-Free Financing:** Unlike traditional credit cards or loans, Tabby's installment plans are interest-free and don't have costly interest charges.

3. **Seamless Integration:** Tabby seamlessly integrates with business owners' existing checkout systems. This allows customers to make smooth payments and grants retailers higher conversion rates.

4. **Risk Management:** Tabby employs super-advanced risk management algorithms and credit scoring models to assess a customer's creditworthiness and reduce the risk of default in loan repayment. This makes for responsible lending practices.

5. **Data Insights:** Tabby provides merchants with valuable data insights and analytics on consumer behavior, purchasing patterns, and transaction performance. This makes them optimize their marketing strategies, inventory management, and pricing decisions.

Since the Tabby platform came into existence in 2019, it has expanded across the MENA region. It has attracted considerable investment from top venture capital firms. It has formed partnerships with several e-commerce businesses, including, but not limited to, fashion houses, electronics manufacturers, travel agencies, and more, thus expanding its reach and customer base.

Tabby has had a transformative impact on the e-commerce style in the Middle East. It has increased online sales, boosted customer engagement, and furthered loyalty for business owners while simultaneously providing customers with greater financial flexibility and convenience. By increasing access to credit money and promoting responsible money-spending habits, Tabby is helping customers make informed buying decisions and improving their shopping experience.

Tabby's innovative platform is changing how people shop online in the Middle East, presenting a win-win solution for consumers and business owners. With its user-friendly interface, transparent pricing, and focus on customer satisfaction, Tabby is leading the pack toward a more inclusive and accessible e-commerce system in the region.

Partnerships:

Tabby's success has been induced by its strategic expansion across the Middle East. The company has steadily increased its presence in several markets, such as the UAE, Saudi Arabia, and Egypt. It uses its host audiences, mainly business owners and customers, to further its adoption and growth. In addition, the platform has partnered with other major e-commerce platforms, payment processors, and financial institutions to extend its offers to a broader audience.

In the UAE, Tabby is a preferred payment option for leading e-commerce players like Noon, Namshi, and Carrefour, enabling shoppers to shop confidently and flexibly.

The company has also collaborated with prominent financial institutions, including Emirates NBD and Mashreq Bank, to offer discounts and promotions to cardholders and build its customer base.

In Saudi Arabia, Tabby has been riding on the wings of the booming e-commerce market to partner with popular virtual retailers like Jarir Bookstore, Xcite, and Extra to offer BNPL options to customers. Tabby has also collaborated with local payment gateways and FinTech start-ups to narrow the checkout process and improve the shopping experience for Saudi consumers.

In Egypt, the platform has tapped into the country's rapidly rising digital economy and has entered alliances with top virtual commerce platforms like Jumia and Souq.

The company also works closely with several local banks and financial institutions to help develop tailored solutions for the Egyptian market and address the needs of customers and businesses.

Looking at the future terrain, Tabby is undoubtedly set for continued growth. It seeks to capitalize on the booming e-commerce market in the Middle East and the evident change in consumer preferences.

The company plans to improve its existing product offers further while expanding its network and gaining entrance into new markets across the Middle East and North Africa (MENA) region. With its love for innovation, customer-centered approach, and collaborations, Tabby is headed for the sky!

Tabby's Middle East Shopping Survey: Unveiling Shopper Insights for Retail Success

Tabby recently conducted the largest shopping survey in the Middle East. This research initiative, which completely covers the field, is set to provide retailers with insights into shopper behaviors and preferences.

In this way, it plans to empower them to optimize their retail strategies for business success. With inputs from more than 7,500 shoppers across the UAE and Saudi Arabia and expert opinions from retail leaders, the platform's survey grants insight into the MENA region's dynamic retail system.

Key Findings

1. Over 30% of shoppers prioritize an excellent shopping experience over brand loyalty. Only 7.7% of consumers mentioned loyalty as a reason for returning to a particular brand to shop.

2. 82% of shoppers reported frustrations with the online shopping process including website navigation, auto-payment options, and long checkout processes. Shoppers aged 18–29 are particularly touchy when it comes to slow e-commerce websites.

3. Female shoppers prioritize affordability over quality more than male shoppers, which shows the importance of pricing strategies to different demographics.

4. 29% of Saudi shoppers expressed frustration when retailers do not offer free shipping, compared to 23% of UAE shoppers, emphasizing the significance of this offering in driving customer satisfaction.

5. 78% of retail customers who shop in a showroom buy items online after seeing them in the physical store to compare prices across different sites. While there's a preference for online shopping, consumers still value expert in-person advice.

6. 60% of buyers abandoned purchases due to payment issues.

Case Study - FAWRY: Digital Finance in Egypt and Beyond

The 'Fawry' Phenomenon of Digital Finance in Egypt and Beyond

Fawry's trajectory from its founding in 2008 to Egypt's first unicorn and largest mobile money platform is simply a result of its visionary administrators, innovative model, and founders' desire to meet the needs of the Egyptian people.

Ashraf Sabry founded Fawry, which has completely transformed how millions of Egyptians manage their finances and conduct money transactions. It provides a more convenient, secure, and accessible platform for various financial services.

Fawry rapidly expanded its network and is now connecting over 35 million people across 166,000 service points all over Egypt. The platform offers more than 850 services, such as utility bill payments, school tuition payments, traffic fines, insurance premiums, and so on. Its extensive reach has made it an indispensable tool for individuals and businesses, enabling them to conduct transactions efficiently.

The success of the platform is most likely due in part to its investments and collaborations. Before the platform went public and got listed on the Egyptian Stock Exchange in 2019, the company had raised $122 million in private equity funding. Its valuation amounted to $1 billion in 2020, a milestone in the Egyptian FinTech industry. Fawry has also been recognized for its technological prowess and innovation, with a Forbes listing as one of the top 25 fintech companies in the Middle East in 2022.

One remarkable fact about Fawry's success is that it is an entirely home-grown scheme. As a result, 100% of its technology is developed by Egyptians.

Fawry is genuinely committed to home-grown innovation and the growth of Egypt's economy. As Egypt's first unicorn, Fawry has set a precedent for future generations of entrepreneurs and tech start-ups as it inspires them to dream big, take risks, and pursue success.

The founder, Ashraf Sabry, while talking about his trajectory in the tech world, stated the importance of being humble enough to realize when one's progress has reached a dead end so one can seek out new opportunities for innovation and growth at that level.

It is remarkable to hear how his previous experiences, especially at Raya, helped shape the concept of Fawry. The realization that investors valued recurring revenue made Sabry explore the payments sector, and his method of seeking out institutional funding rather than following the traditional start-up pathway shows innovative thinking and confidence in his vision. By securing funding from banks, private equity firms, and corporate organizations, he laid a solid foundation for Fawry's success.

Fawry's success can be attributed to several *key factors:*

1. ***Convenience:*** Fawry has made it convenient for its users to access services through diverse means, such as mobile apps, online platforms, ATMs, retail outlets, and kiosks. Fawry users can make payments or access financial services anytime and anywhere using their preferred method.
2. ***Accessibility:*** Fawry has many partners across banks, financial institutions, government agencies, e-commerce platforms, retailers, and service providers.
3. ***Innovation:*** Fawry loves innovation and continuously invests in developing new tech solutions to meet users' and businesses' dynamic and insatiable needs. The platform regularly

introduces new features and upgrades its infrastructure to stay ahead in fintech.

4. ***Security:*** The platform prioritizes security and reliability, using its encryption technologies, authentication mechanisms, and fraud detection systems to protect its customers' personal information and gain their continued trust.

5. ***Customer-centred:*** Fawry prioritizes customer satisfaction and strives to deliver exceptional service at every point.

6. ***Financial Inclusion:*** Fawry provides access to essential financial services for individuals who may not have access to traditional banking channels.

7. ***Merchant Solutions:*** Fawry offers several merchant solutions that help businesses narrow their operations, increase efficiency, and grow their customer base.

8. ***E-commerce Integration:*** Fawry is in sync with several e-commerce platforms and online markets to provide payment solutions for merchants and customers.

9. ***Financial Education:*** Fawry also promotes financial literacy and empowerment among its users by offering educational resources and training workshops to help individuals and businesses understand financial concepts, manage their money effectively, and make knowledgeable financial decisions.

10. ***Social Impact Initiatives:*** the company supports social impact initiatives and community development projects across Egypt. It collaborates with nonprofits, NGOs, and the government to address pressing social issues and SDGs.

Expansion

Fawry always has alliances with top contenders in different fields, such as retail, telecommunications, utilities, etc. These partnerships help the company offer many more financial services to many more people.

Through its host of partnerships, Fawry expands its offerings and audience and continues delivering valuable tech solutions to empower businesses and contribute its quota to the growth of Egypt's digital economy.

Fawry has announced collaborations with other top companies in different sectors. These partnerships are at the core of Fawry's mission of expanding its reach and increasing accessibility to digitalized financial services for people and organizations.

One striking collaboration is with Edita, a top food industry player. This collaboration aims to develop collection and payment systems within the food industry sector. It is a step forward in Fawry's trajectory in the Egyptian market, increasing convenience and digital accessibility.

Case Study - OPTASIA: Navigating the Future of Fintech Brilliance

Optasia – AI-powered Digital Solutions

Optasia was founded in 2012 by Bassim Haidar. Formerly known as Channel VAS, it is a leading financial service provider with its headquarters in the UAE. Optasia offers a wide range of financial services, including airtime services, micro-lending, and data monetization solutions, but they prioritize services to mobile operators and financial institutions. Optasia uses AI technology to serve a broad customer base spanning over 30 countries across Sub-Saharan Africa, the Middle East, Asia, and Latin America. The company focuses on emerging markets and has provided accessible financial services to underserved audiences for over a decade.

The company has achieved notable milestones with over 45 active deployments and $40 billion in credit decisions. Optasia has disbursed over $10 billion in advance funds between 2019 and 2022, with $3.5 billion disbursed in 2022 alone. The platform had a monthly user average of 95 million in 2022, showing its widespread adoption and impact. Optasia also has a sizable investor interest with Promoter Holding, Chronos Capital Limited, Waha Capital, and Ethos. In 2022, the company's valuation surpassed the $1 billion mark, putting paid to the argument that it is now a prominent player in the fintech industry.

The company has played an active role in the mobile and financial revolution in State economies across Africa, the Middle East, Asia, Europe, and Latin America. Optasia has established itself as a trusted collaborator for several global names. The company's credibility has earned it further recognition from other firms. Optasia also attained ISO 27001 Certification in 2021, which shows the high standards of its information management system. Two years ago, Optasia began a rebranding journey and adopted a new identity that aligns with its goals and aspirations. The name "Optasia" is Greek for "vision," and it captures the platform's commitment to innovation and expansion.

Optasia has the following characteristics that set it apart from other FinTech solutions:

- **Data and AI Technology:** Optasia uses advanced data analytics and AI technologies to get insights, identify trends, and develop bespoke solutions.
- **Financial Services:** The company offers a wide range of financial services, including mobile financial services, microloans, digital banking solutions, and other financial products, all aimed at improving access to financial resources and inducing economic growth.
- **Partnerships:** Optasia partners with international brands, financial institutions, telecommunication providers, and other stakeholders to expand its reach and leverage existing networks and infrastructure to deliver more effective services.
- **Business Development:** Optasia jumps on opportunities for growth by identifying emerging markets, assessing market demand, and developing strategies to enter new territories. The company continuously adapts its business model to capitalize on evolving trends.
- **Regulatory Compliance:** Optasia focuses on compliance with legal regulations in the jurisdictions in which it operates. This includes obtaining all the required certifications, considering data privacy, and adhering to industry standards and global best practices.

Benefits

1. **Decreased Cost of Lending:** Optasia's tech solutions streamline lending processes by automating repetitive tasks and decreasing the resort to manual methods. They also help lenders reduce operational costs for loans and related services.

2. **Exacerbated Efficiency:** Optasia enables lenders to make more informed lending decisions using advanced algorithms and data analytics, thus minimizing potential losses. This enhanced risk management leads to increased efficiency and lower costs.

3. **New Services and Revenue Streams:** Optasia provides innovative tools and expertise to help financial institutions develop novel bespoke financial services. These new solutions help businesses generate more revenue and diversify their offers.

4. **Alignment with Trends:** Optasia assists businesses in identifying untapped opportunities for innovation, and by staying ahead of industry trends and customer preferences, clients can capitalize on emerging markets and remain competitive in this dynamic financial environment.

5. **Access to New Markets:** Optasia offers access to an extensive network of partners comprising financial institutions, telecommunications providers, and technology companies spanning various regions and industries. Businesses can easily break into new markets and reach previously untapped areas.

6. **Market Expansion:** Whether entering new markets in new locations, targeting underserved audiences, or diversifying offers, Optasia offers more than enough support and expertise to facilitate market expansion.

7. **Micro and Nano Lending Solutions:** Optasia's portfolio consists of both micro and nano lending solutions, which provide quick and convenient access to small loans for

individuals and SMEs. These solutions can be easily integrated into other financial institutions, mobile network operators, and other stakeholders to offer microloans to their customers.

8. **Airtime Advance:** Optasia offers flexible airtime, data, and package credit solutions for MNOs, enabling them to provide convenient airtime advances to their customers. This improves customer satisfaction while also generating additional revenue streams for MNOs. Through its solutions, MNOs can improve customer retention, attract new subscribers, and differentiate themselves in competitive markets.

9. **Data Monetization:** Optasia enables MNOs to unlock the value of their data and create new revenue streams. By increasing the value of their data assets, MNOs can boost their competitiveness and revenue and increase customer engagement.

Partnerships

Recently, Optasia announced that JS Bank in Pakistan is leveraging its tech-driven platform to drive its micro-lending solution as part of its digital banking plan.

Using the commercial name 'Zindigi,' JS Bank's digital banking initiative in Pakistan offers a wide array of offers in its digital banking app: digital loans, remittances, digital payments, stock trading, mutual funds, and premium debit card services.

Zindagi has also established partnerships to expand its offerings as a Banking as a Service (BaaS) and open banking solution provider. The integration of Optasia's AI platform is set to enhance and optimize Zindigi's mobile financial services, facilitating financial inclusion and economic development while benefiting underbanked customers.

Mark Muller, Optasia's Group CEO, discussed the significance of the company's AI-led platform in enhancing global services, stating, "The capabilities and robustness of Optasia's AI-led platform are valued every day by partners across the globe that seek to improve the services they offer."

Noman Azhar, Chief Officer of Zindigi, stressed the critical role of Optasia's AI capabilities in optimizing digital banking services and making advanced lending accessible to all, stating, "At Zindigi, our commitment is resolute to catalyzing a positive transformation within the financial services landscape, ultimately benefiting society as a whole."

With a presence spanning over 30 countries, Optasia's proprietary AI-led platform enables millions of unbanked individuals and SMEs instant access to financial solutions. The company's B2B2X model creates value for partners such as mobile network operators, banks, and payment gateways without additional operating or capital expenses.

Case Study - Paytabs: Pioneering Seamless Payment Experiences

PayTabs -Payment Orchestration Solutions

PayTabs was founded in 2014, and its headquarters are in Saudi Arabia. Since then, the company has become a payment solutions pioneer, digitally transforming commerce experiences for merchants and customers across the Middle East.

Abdulaziz Al Jouf's group founded it. The platform leverages proprietary technology to offer users an end-to-end suite of innovative payment products, thus facilitating business transactions such as digital invoicing, QR code payments, social media commerce, point-of-sale systems, and unified payment.

PayTabs serves over 1.1 million active business persons in various industries across Saudi Arabia, the UAE, Egypt, Oman, Kuwait, Jordan, Palestine, Türkiye, and Azerbaijan.

In 2022, PayTabs launched its pioneering PayTabs SwitchOn solution, consolidating payment processing for businesses across diverse global payment types through one unified interface. That same year, the acquisition of Digital Pay in Saudi and Turkish social commerce platforms Paymes expanded PayTabs' reach in the Arab market.

Today, the company has over 225,000 mobile app downloads, a testament to its convenient digital payments proposition. In 2022, the total payment volume powered by PayTabs reached an all-time high of $8.2 billion. Despite all this company has achieved, it still seems to be soaring even higher.

Story

PayTabs was initially conceived as a "napkin idea" by its founder to tackle the complicated world of virtual payment processing he experienced during an entrepreneurial venture in the past. He realized that most small businesses faced payment system challenges and considered a better solution. The result was PayTabs - an early pioneer in the then-little-known financial technology space that has become one of the fastest-growing financial sectors. As a payment gateway and technology provider, PayTabs integrates smartphones and traditional payment channels like credit cards, providing smooth payment options.

PayTabs was created to provide easy access to digital payment infrastructure for merchants locked out of systems; PayTabs is a solution for overcoming e-commerce and digital payment growth challenges in the Middle East. The innovative idea gained seed funding from Saudi Aramco's Wa'ed venture fund to kickstart the company. Over time, PayTabs successfully tackled payment obstacles through technological advancements, showing the potential of fintech and driving the expansion of regional e-commerce.

The founder's concept of democratizing payments for underserved SMEs has established PayTabs as a pioneer in the payments industry.

PayTabs offers its clients a simple plug-and-play payment solution. Other features, like the complimentary Smart Business Manager, empower business people to monitor sales, develop customer retention strategies, and create loyalty programs at no extra cost.

Through technological innovation and adaptability to evolving client needs, PayTabs has expanded well beyond a transaction processor to become an enabler of digital commerce and financial access. The company swiftly onboards merchants for e-payments and continues improving its offers while pioneering cashless adoption across the Middle East.

PayTabs operates on a model that provides payment solutions to businesses across different sectors. Here's a breakdown of PayTabs' business model:

1. Payment Processing Services: PayTabs is a payment aggregator, facilitating secure and efficient online payment processing for businesses with various payment methods, including credit/debit card payments, bank transfers, and alternative payment options for diverse customer preferences.

2. Subscription-Based Revenue: PayTabs generates revenue through subscription-based models where businesses pay a recurring fee to access its payment processing platform. This model ensures a steady income stream and allows businesses to budget effectively for payment processing services.

3. Transaction Fees: PayTabs charges a transaction fee for each payment processed through its platform. This fee varies based on factors such as transaction volume, payment method used, and geographic location. The transaction fee model allows PayTabs to generate revenue based on actual platform usage.

4. Value-Added Services: PayTabs offers value-added services like fraud detection and prevention, data analytics, and reporting to help businesses optimize payment processes and promote security. These services may be offered as add-ons for an additional fee, providing a revenue stream for PayTabs.

5. Cross-Border Payments: PayTabs facilitates cross-border payments so businesses can accept payments from customers in different countries. It only charges a currency conversion fee for such international transactions.

6. Partner Integrations: PayTabs partners with other companies like e-commerce platforms, website builders, and financial institutions to integrate its payment processing services.

PayTabs can expand its customer base through these partnerships and generate revenue through referral fees or revenue-sharing agreements.

7. Merchant Onboarding and Support: PayTabs charges a one-time setup fee or provides support services to business owners who join the platform.

Benefits of PayTabs Invoices

PayTabs Invoices is an electronic invoicing system that enables businesses to manage, send, and receive payments via digital channels. Users can generate customized, itemized invoices and accept instant payments via QR codes, emails, SMS, and social platforms while tracking finances in one place. The features include multi-currency support per product listings, QR codes for simplified payment redirects without additional apps, theme customization options, and consolidated invoice management for multiple customers on a single dashboard.

By shifting from manual processes to PayTabs Invoices, businesses can execute transactions faster while growing through simplified invoice sharing and digital payment collection. The solution also eases financial control and reconciliation through organized reporting on paid invoices. For sellers, the system surfaces invoice critical details via QR codes to buyers, including seller name, tax ID numbers, VAT applicability, and totals payable. Additionally, businesses can download statements on all invoices settled by customers.

PayLinks

In today's world of social commerce, PayTabs simplifies payments with PayLinks. PayLinks are secure web links you can share easily through digital or social platforms. Whenever you share a PayLink, your customers can click on the link, which will direct them to a secure payment page. When they arrive, they can complete their transaction on that page, and you will receive payment instantly. You can use PayLinks across different social media platforms like Facebook, Instagram, Twitter, LinkedIn, and Pinterest, as well as emails. The beauty of PayLinks is its simplicity - they are just links you can use to increase sales and income on your social media posts, website, or any digital channel of your choice.

Benefits of PayLinks

1. PayLinks provides a convenient payment solution for creating promotions instantly on any social or digital platform.
2. For existing business owners, creating PayLinks is easy. Log on to the PayTabs Merchant Dashboard, go to the PayLinks section, and follow the instructions on the screen to generate your next PayLink.
3. Getting started with PayLinks is quick and easy for new customers. Sign up first, then access the PayTabs Merchant Dashboard, select PayLinks, and follow the on-screen instructions to create your next PayLink.
4. With PayLinks, you can streamline your payment process, enhance your online presence, and induce sales across multiple digital platforms.

Repeat Billing with PayTabs

Using PayTabs' repeat billing capabilities, you can streamline subscriptions, improve your convenience with digitalized payments, and grow your business. PayTabs repeat billing solutions are designed to meet companies' diverse needs, offering exciting features and simple integration to support subscription-based business models.

Partnerships

The PayTabs Group recently partnered with the BNPL platform Tabby to support the growth of e-commerce across Saudi Arabia and the UAE. They intend to do this by attributing Tabby's interest-free split payment style to the PayTabs business owners' network. The partnership aims to transform regional e-commerce by offering a stress-free 'Buy Now, Pay Later' experience.

The partnership helps business owners who use PayTabs to offer Tabby's flexible payment options on their checkout pages. This connotes that shoppers can now split their purchases into four interest-free installments. Invariably, this translates to higher cart values and an expanded customer base for retailers. In reality, Tabby has witnessed higher demand for its offers, which is in tandem with the e-commerce boom in the UAE and Saudi Arabia. The boom is being furthered by the adoption of digitalization and the shifting millennial shopping preferences, especially in categories like fashion, electronics, and furniture.

PayTabs partnerships are structured across three levels to foster collaboration and mutual benefit:

- Strategic Partnerships: Strategic partnerships are established to align with common business objectives and goals. These partnerships are entered into to achieve long-term objectives like expanding market reach, improving product offerings, or entering new markets. PayTabs plans to gain enough complementary strength and resources through these partnerships to further sustainable growth and innovation.

- Operational Partnerships: Operational partnerships are formed to provide value-added services to customers or enhance operational efficiencies. These partnerships combine complementary services or technologies to deliver all-encompassing solutions to customers. They may involve joint product development, co-marketing efforts, or shared resources to optimize processes and increase customer satisfaction.

- Resource Partnerships: Resource partnerships are established to leverage human, technical, or information resources to mutual advantage. These partnerships may involve collaboration on research and development initiatives, knowledge sharing, or access to specialized expertise. Resource partnerships enable PayTabs to tap into additional capabilities or expertise that may not be available internally, facilitating innovation and driving competitive advantage.

By collaborating across these three tiers, PayTabs seeks to create a network of partners who collectively contribute to its growth and success. Partnerships play an essential role in improving PayTabs's capabilities, expanding its market presence, and adding value to customers.

Case Study - TAMARA: Shop and Split Your Payments

Tamara - Flexible Payments

Tamara, a Sharia-compliant BNPL service, has been a major player in the Arab market since its establishment in 2020. It was founded in Riyadh, Saudi Arabia, by serial entrepreneurs - Abdulmajeed Alsukhan, Turki Bin Zarah, and Abdulmohsen Al Babtain.

Tamara has proved to be a force to contend with in the FinTech world, emphasizing customer satisfaction, transparency, and adherence to ethical values. With its headquarters in Saudi Arabia, the company has gained much traction in FinTech, significantly when it raised almost $116 million across four funding rounds.

Investors in Tamara include Impact46, CheckOut.com, and Nama Ventures, among others. This support shows confidence in Tamara's innovative approach to revolutionizing the payment experience for both business owners and consumers.

Tamara's primary goal is to create a seamless shopping experience by offering zero-interest fees for its services. By providing flexible payment styles, Tamara helps to increase affordability and convenience for its customers while boosting sales and income for business owners.

In 2021, Tamara made headlines with its record-breaking Series A funding round, raising $110 million. This substantial investment shows the growing demand for buy-now-pay-later solutions, strengthening Tamara's stance as a top player globally in the FinTech world. Abdulmajeed Alsukhan's observation of the debt ledger system in local neighborhood shops inspired the idea for Tamara. The system allowed customers to buy items on credit and settle accounts at the end of the month.

This concept inspired Abdulmajeed, and after researching global markets, he decided to do something similar and introduce another BNPL service to the Arab market. This was what led to the establishment of Tamara in Riyadh. It was created to fill a gap in the market, and so far, it has done precisely that.

Unlike other traditional BNPL models, Tamara's business model involves taking a percentage of the purchase price from the stores rather than the customers. This approach encourages customers to make purchases beyond their immediate budget, and in the long run, both consumers and business owners will benefit from the arrangement.

The company's Sharia-compliant nature means it does not charge interest on payments, which makes Tamara stand out in the FinTech market. Tamara's commitment to Sharia ethical practices is seen through its refusal to charge interest on payments and its charitable contributions from late fees. By placing fairness, transparency, and social responsibility at the helm of its product offerings, Tamara builds trust and loyalty among its customer base.

Tamara has experienced tremendous growth since entering the FinTech space. So far, it has expanded its operations worldwide, with several offices in multiple countries. The company's $110 million Series A funding round in 2021, led by Checkout.com, was a record-breaking milestone in the Middle East's fintech history.

Benefits

Tamara's Several benefits stand out in the buy now, pay later (BNPL) market. Some of them are explained below:

1. Sharia Compliance: Tamara operates on Sharia-compliant principles. This is why it does not charge interest on payments.

This aligns with Islamic finance principles and makes the platform appeal to a wide range of customers searching for ethical financial solutions, mainly Muslim customers.

2. Zero-Interest Instalments: Customers can make purchases and pay for them in interest-free installments over some time. This option connotes flexibility and affordability, allowing customers to transfer payments without additional costs.

3. Transparent Pricing: Tamara's pricing structure is transparent, with no hidden fees or charges. Customers know the total amount they need to repay upfront, which promotes trust and confidence in the platform.

4. Seamless Checkout Experience: The platform offers a stress-free checkout experience for customers, allowing them to complete purchases quickly and easily. Since Tamara is linked to many online stores, customers can select Tamara as a payment option at checkout, streamlining the payment process.

5. Instant Approval: Customers can instantly get approval for Tamara's BNPL service, which helps them make purchases without delay. This convenient feature allows customers to take advantage of promotional offers or discounts in real-time.

6. Flexible Repayment Options: Tamara provides flexible repayment options and allows customers to choose from various installment plans based on their choices and financial circumstances. Customers can select the number of installments they want and the repayment plan that best suits their needs.

7. Charitable Contributions: Tamara donates a portion of customer late fees to charitable causes, demonstrating its commitment to social responsibility and giving back to the community.

8. Partner Merchant Network: Tamara collaborates with several merchants, including global brands and local businesses, and

offers users a range of products and services. This extensive network improves customer choice and accessibility.

Partnerships

A prominent financial services enabler in the Arab region - Paymob, recently partnered with Tamara, an innovative Buy Now Pay Later (BNPL) platform in the GCC area. The plan is to transform Tamara's BNPL service into Paymob's secure gateway to facilitate smooth payments and assist customers in splitting their payments without hidden fees or interest rates.

Islam Shawky, the co-founder and CEO of Paymob, stressed the significance of collaboration in inducing the growth of SMEs in the digital economy. By leveraging alternative payment methods, the alliance with Tamara is an excellent opportunity for business owners in the GCC area to expand their capabilities by aligning with Paymob's mission to drive growth in the MENA region.

What is significant is the fact that this collaboration between two of the region's fastest-growing FinTech institutions will create a payment system that will provide business owners with comprehensive solutions and seamless customer experiences. The primary focus of the partnership, though, is to spur SME growth and contribute significantly to the GDP in the MENA region. Businesses of all sizes stand to benefit from increased sales and conversions through its comprehensive payment gateways.

Turki Bin Zarah, the co-founder and CCO of Tamara, talked about the role of the partnership in granting thousands of SMEs seamless access to Tamara's services, thereby facilitating their growth across the region. Integrating Tamara's BNPL solution into Paymob's gateway through a simple integration removes payment hurdles and makes for seamless, secure transactions.

Other Notable FinTech Startups in the Arab region are:

Paymob (Founders - Islam Shawky, Alain El-Hajj, Mostafa Menessy (Egypt))

PostPay (Founder - Tariq Sheikh (UAE))

HyperPay (Founder - Muhannad Ebwini (Saudi Arabia))

Khazna (Founder - Omar Saleh, Ahmed Wagueeh, Fatma Shenawy (Egypt))

CHAPTER 2

Unveiling the Future: Cutting-Edge Tech Transforming E-commerce

Not long ago, anyone searching for tech-powered eCommerce moved to the West. However, the Middle East seemed to have taken over. The Arab world is uniquely advantaged with its young and tech-savvy audience, rapid technological advancement, rising disposable incomes, increased internet access, and changing consumer preferences. Most Arab countries spark lucrative opportunities for businesses seeking to tap into the booming online market.

1. Booming eCommerce Market

The Arab eCommerce market has grown exponentially in recent times, and this fact is due to factors such as enhanced smartphone penetration, increased internet access, and a growing middle class with improved purchasing power. According to several reports, the eCommerce market in the Middle East and North Africa (MENA) region is projected to reach $28.5 billion by 2022, showing over $28 billion growth potential.

2. Young Population

In all Arab countries, the young and tech-savvy population of primarily millennials and Gen Z are now so much in tune with online shopping, and this is furthering the growth of eCommerce in the region. Any business that caters to their preferences and stands to benefit tremendously from the rising demand for online shopping.

3. Increased Internet Access

Recently, internet connectivity in the Middle East has been astronomically high due to improved infrastructure, smartphone affordability, and government support. This has further increased the prevalence of online shopping, thus allowing businesses to reach a larger audience and expand their customer base beyond geographical limits.

4. Mobile Commerce

Many online transactions in the Arab region are now conducted through smartphones and mobile devices. This is the reason why mobile commerce, or m-commerce, is on the rise. To further this end, Arab countries are optimizing eCommerce platforms for all mobile users and offering seamless mobile shopping experiences.

5. Government Support

Governments in Arab countries have realized the vast potential of investment in tech, specifically in eCommerce, and have recognized eCommerce as an economic growth and job creation driver. As a result, many governments have implemented several policies to support the development of the digital economy and eCommerce sector. These policy programs include investment in digital infrastructure, support for digital entrepreneurship, and regulatory legal reforms to facilitate online business operations.

Improving eCommerce Margins

The evaluation proffers four strategies that can enhance margins in eCommerce:

1. Increased Transparency:

The current eCommerce profit and loss margin lines are so blurred that it is difficult to determine which is which. There should be transparency in this area so business leaders can make informed decisions. This would most likely connote the imputation of eCommerce metrics into business reporting systems. When business leaders comprehend the profitability of online business interactions, they can then allocate resources effectively, make informed investment decisions, and optimize margins.

2. Specific Marketing:

Most businesses should earmark specific investments for eCommerce marketing rather than drawing marketing budgets from greater shopper-marketing allocations. Resources are dedicated to driving online sales and maximizing ROI in the digital space. Thus, businesses can effectively target online consumers, optimize conversion rates, and increase revenue growth by prioritizing eCommerce solutions.

3. Revenue Growth Management Tactics:

Businesses can use revenue growth management tactics that align with eCommerce channels to boost growth. They can do this by introducing channel-specific products to prevent consumers from directly comparing prices and by offering unique products through eCommerce channels. These actions differentiate them from competitors and help them retain their pricing power.

4. Optimize Omnichannel Supply Chain Actions:

Incorporating omnichannel supply chain actions is also important as it helps the eCommerce platform optimize margins. It involves improving demand forecasting accuracy, enhancing execution precision, and redesigning packaging to mitigate costs.

Case Study - SARY: Empowering Businesses, Enriching Communities

SARY- Online Wholesale Market

Sary is a Saudi-Arabian B2B e-commerce platform that connects small businesses. It has recently risen to prominence in the Arab digital marketplace. Sary was founded in Riyadh, and it primarily caters to small and medium-scale businesses (SMBs). The organization is committed to addressing the unique needs of businesses in Saudi Arabia. So far, it has established itself as a major player in the country's growing eCommerce sector.

Key Features:

1. B2B Marketplace: Sary is a comprehensive B2B marketplace that connects suppliers, wholesalers, retailers, and other businesses across a plethora of industries. The platform offers a wide range of products, including but not limited to groceries, household items, and electronics, catering to the diverse needs of businesses in Saudi Arabia.

2. Convenience and Efficiency: Sary provides businesses with a convenient and efficient platform to procure products and manage their processes. It has a user-friendly interface and several intuitive functions that businesses can use to browse, compare, and purchase products seamlessly, saving time and streamlining operations.

3. Wide Range of Products: One of Sary's key strengths is its vast catalog of products sourced from trusted suppliers and brands. From everyday necessities to specialized items, businesses can find everything on the platform, eliminating the need to visit multiple suppliers or wholesalers.

4. Competitive Pricing and Deals: Sary offers competitive prices on many products, allowing businesses to optimize procurement costs. The platform usually offers discounts,

promotions, and bulk-buying options that enable businesses to access cost-effective solutions for their buying needs.

5. Delivery and Logistics: The platform also provides reliable delivery and logistics services, which ensure timely and efficient order fulfillment. Businesses can schedule deliveries based on their specifications, with options for both standard and express shipping.

6. Dedicated Customer Support: Whether it's resolving inquiries, handling returns or exchanges, or providing assistance with orders, the platform's customer support team is available to address business needs promptly and efficiently.

Business Model

Sary operates a business-to-business (B2B) eCommerce model as it is an all-encompassing marketplace connecting suppliers and wholesalers with retailers and other businesses in Saudi Arabia. This is a breakdown of Sary's business model:

a) Supplier and Wholesaler Partnerships: Sary has a network of suppliers and wholesalers, including manufacturers and distributors, all with one task or the other on the chain that helps source products for its platform. Partnerships with these different brands help Sary offer a varying range catalog of goods and products, thus ensuring that retailers can access a wide selection of items to meet their customers' needs.

b) Online Marketplace: Sary operates an online marketplace that avoids a physical meeting point. Its platform features an intuitive interface and search functionality that allow users to find the items they need and seamlessly place orders.

c) Revenue Generation: Sary generates revenue through several channels

d) Commission Fees: Sary charges suppliers and wholesalers commission fees for each transaction facilitated through its platform. These fees may be a percentage of the total order value or a fixed amount per transaction.

e) Subscription Fees: Sometimes, Sary offers subscription-based plans to businesses, which provides them with access to premium features, personalized services, or increased support. The subscription fees contribute to the platform's recurring revenue stream.

f) Advertising and Promotions: Sary may offer advertising opportunities to suppliers and wholesalers, allowing them to promote their products or brands on the platform. Advertising fees and promotional placements generate additional revenue for the business.

g) Delivery and Logistics: at Sary, there are delivery and logistics services to ensure the timely and efficient fulfillment of orders. Sary may charge delivery fees to businesses based on factors like order size, delivery distance, and service level. They may partner with third-party logistics providers to improve delivery operations and expand across Saudi Arabia.

h) Value-added Services: In addition to being a core eCommerce platform, Sary offers value-added services to business organizations, such as inventory management tools, analytics dashboards, and procurement optimization solutions. These services help businesses narrow down their operations, boost efficiency, and make informed decisions about their purchasing strategies.

Partnerships:

These play essential roles in Sary's development and growth since it seeks to solidify its position as a top eCommerce platform. There are a couple of ways Sary can approach expansion, and they are:

- Geographic Expansion: Sary can explore opportunities to expand its operations beyond the Saudi Arabian market. This will involve entering new territories within the Gulf Cooperation Council (GCC) countries like the United Arab Emirates, Qatar, Bahrain, Oman, and Kuwait. This is because there is a growing need for B2B eCommerce solutions in these locations. Expansion into neighboring markets will allow Sary to capitalize on the region's evolving business landscape.

- Product Expansion: Sary can diversify its product portfolio to cover broader business needs. New categories like office supplies, healthcare products, household items, groceries, and construction materials can be included. By expanding its product portfolio, Sary is increasing its value proposition and attracting more business.

- Service Expansion: Sary can enhance its service offerings to provide additional value to its customers. This may include offering premium membership plans with exclusive benefits, such as faster delivery times, dedicated account managers, and personalized support. Add Sary can also introduce value-added services such as inventory management solutions, bulk ordering discounts, and customized procurement solutions tailored to the needs of large enterprises.

Partnership Strategies

- Supplier Partnerships: Sary can ensure a steady stream of product supply on its platform by collaborating with top brands and manufacturers, enabling Sary to offer exclusive products and competitive pricing.

- Logistics Partnerships: Sary can form partnerships with delivery providers to optimize its operations and improve the efficiency of its supply chain. By partnering with reputable logistics companies, Sary can offer reliable and timely delivery services to businesses across its target markets. Collaboration in this manner will also mean reaching out to remote areas and expanding its customer base.

- Collaborations: Sary can leverage technological collaborations to improve its platform's functionality and user experience. This can be done via partnerships with software providers to impute advanced characteristics like AI-driven product recommendations, predictive analytics, and inventory optimization tools. Seeing Sary improve its trajectory and crystallize its position as a top B2B eCommerce platform would be great.

Case Study - MaxAB: Redefining Excellence in eCommerce

MaxAB - e-commerce platform connecting wholesale food and grocery

One start-up that has turned the traditional grocery supply chain in Egypt and beyond around is MaxAB. It was established in Egypt in 2018 by Belal El-Megharbel and Mohamed Ben Halim. Since then, it has proven itself to be one of the top leaders in the lucrative e-commerce and logistics sector in the Middle East and North Africa. The organization operates as a virtual B2B market that links up informal food and grocery retailers with suppliers through its e-platform.

Business Model:

1. Digital Market: MaxAB has a virtual market that is a one-stop shop for retailers to buy different kinds of food and groceries. The customers can search through different product catalogs, place their orders, and manage their inventory via its user-friendly website and app.
2. Supplier Network: The start-up is in league with several suppliers, including manufacturers, wholesalers, and distributors, all of whom source products for its platform. The MaxAB platform uses technology and data analytics to simplify the sourcing process and ensure that retailers have access to high-quality products at competitive prices.
3. Delivery: The platform also delivers products to consumers, streamlining the supply chain. It has its own fleet of delivery vehicles and a team of drivers on its staff, ensuring timely and efficient delivery of orders to retailers' doorsteps.
4. Tech-driven Solutions: The MaxAB platform uses advanced technology, such as data analytics, machine learning, and suggestive algorithms, to forecast, manage inventories, and optimize routes. These technologies help the start-up reduce

delivery times, decrease stock-outs, and improve operational efficiency.

Features

1. Variety of Products: The MaxAB platform offers a variety of food and groceries, including fresh produce, dry goods, beverages, and household necessities. Their customers, as well as retailers, have access to different catalogs and can easily meet all their inventory needs.

2. Digital Platform: The MaxAB platform has a very user-friendly digital interface, whether on the site or the mobile app. This interface allows retailers to order, manage their inventory, and track deliveries everywhere, every time.

3. Supplier Network: The MaxAB start-up collaborates with a network of trusted suppliers, which ensures a reliable and consistent supply of goods to be sold. Retail customers can access high-quality products at competitive prices through MaxAB's extensive supplier network.

4. Inventory Management Tools: MaxAB allows its customers to use advanced inventory management tools powered by data analytics and machine learning algorithms. These tools help retailers reduce stockouts, optimize their inventory, and improve store efficiency.

5. Predictive Analytics: The MaxAB start-up usually uses suggestive analytics to forecast customer demand and anticipate market trends. This informs retailers beforehand and guides them on their buying strategy, profit increase, and wastage mitigation.

6. Timely Delivery: MaxAB has its fleet of delivery vehicles with a team of drivers. This helps them deliver efficiently and on time to retailers. Consequently, retailers can rely on them for prompt

delivery services, helping them maintain uninterrupted operations.

7. Cost Reduction: By getting their products straight from the suppliers and optimizing the supply chain, the platform helps retailers minimize the cost of procurement. These retailers benefit from competitive pricing, bulk-buying discounts, and straight logistics, which results in cost savings over time.

8. Convenience: The MaxAB platform offers convenience and efficiency to retailers by providing a stress-free procurement exercise. With its digital platform, retailers can easily place orders, track deliveries, and manage inventory, thus saving time and effort.

9. Customer Support: MaxAB proffers customer support to retailers who need assistance via answers to questions on issues or concerns they may have. Their customer support team is quite responsive, and they are there to ensure retailers have a positive experience at every step.

Expansion Strategies:

1. Geographic Expansion: recently, MaxAB started an expansion strategy that is intended to scale its operations across Egypt and beyond. The start-up has extended its operations to retailers in major cities across the country, including Cairo, Alexandria, and Giza. MaxAB also plans to explore opportunities for international expansion so it can tap into new markets in the Middle East and Africa.

2. Product Diversification: MaxAB is currently diversifying its products to meet retailers' dynamic needs. The organization continuously expands its catalog to include more food and groceries, such as fresh produce, dry goods, beverages, and household necessities.

Partnership Strategies:

1. Supplier Partnerships: MaxAB works with several suppliers to obtain more products and maintain a steady supply of goods. The company has a penchant for partnerships with trusted suppliers and brands to offer retailers access to high-quality products at competitive prices.
2. Logistics Partnerships: MaxAB also links with logistics and delivery service providers to improve its delivery capabilities.

Merger with Wasoko

The Wasoko and MaxAB merger is quite an essential milestone in the evolution of the Arab region's e-commerce scene. The partnership has co-joined two of the continent's top B2B players in the e-commerce space with a shared vision of driving growth and development in Africa's informal retail sector.

1. Market Leadership and Scale: The combination of both will emerge as the largest tech merger in Africa. Both have a customer base of over 450,000 business owners serving an estimated 65+ million consumers across eight African countries. This positions the merged company as a formidable force in Africa's e-commerce sector with the potential to impact the retail scene.
2. Sustained Growth and Traction: Both Wasoko and MaxAB have impressive growth trajectories and enough traction in their respective markets. While Wasoko's monthly revenue has increased by 30%, MaxAB has grown its monthly active business network by 25% and fintech transaction volumes by more than 50%. This shows that there is a strong demand for their products in the market and indicates future growth prospects.

3. Pan-African Vision: the merger also shows a shared pan-African vision to provide solutions to major challenges that are barring the development of Africa's informal retail sector, which is presently estimated to be worth $850 billion.

4. Integrated Solutions: Both Wasoko and MaxAB are known for their integrated payment solutions, merchant financing, and proprietary logistics operations, all of which form the cornerstone of their businesses. Having a merger automatically connotes having a combined entity strengthen its infrastructure and harness cross-market synergies that will drive operational efficiency and make for a smooth customer experience.

5. Executive Leadership and Commitment: Belal El-Megharbel, CEO of MaxAB, and Daniel Yu, CEO of Wasoko, are both committed to shaping the long-term future of the merged company. They will continue as full-time executive leaders while bringing their expertise, vision, and leadership to boost the company's growth in the years ahead.

6. Investment and Runway: as part of the merger, the combined business has received additional investment and has a substantial runway to reach profitability while pursuing new opportunities. This infusion of capital will sponsor expansion, innovation, and wonderful initiatives that will further solidify the company's position as a market leader in Africa's e-commerce scene.

Case Study -BRIMORE: An Omnichannel Social Commerce Platform

Brimore - Social Commerce Platform connecting Emerging Brand Owners

Brimore is a start-up founded by Mohamed Abdulaziz and Ahmed Sheikha to solve a nutty issue in the Egyptian market—the struggle of new brands to succeed despite the crowd of price-sensitive customers. The primary hindrances these brands face are marketing and distribution, which usually require a great deal of finances. Brimore wants to solve this problem by connecting new brand owners to a network of micro-sellers on social media via its social e-commerce platform.

This will empower these micro-sellers to promote and sell products within their social circles, which invariably implies that Brimore creates and generates demand for unknown brands. This approach makes for smooth market access for suppliers and provides sellers with diverse income opportunities.

The platform's name, Brimore, shows its mission to "bring more" opportunities for new brands and micro-sellers. This approach rides on the rapid growth of e-commerce in the Arab region, as the sector witnessed a 52% surge to $22 billion at the end of 2020, which was major due to the Egyptian, Saudi Arabian, and UAE markets.

Mohamed Abdulaziz and Ahmed Sheikha shared a background in end-to-end distribution and product management. Together, they established Brimore in 2017, and even though they started with just five employees, the company has grown so much that its team has increased to about 700 individuals.

Brimore has a network of more than 75,000 active sellers and 300 suppliers. It has more than 8,000 products across diverse groups, such as household goods, personal care, furniture, fashion, food, electronics, and beverages.

The plan is to unravel the Egyptian market's potential and induce more growth in the e-commerce field. Their mission increases the demand for the products of new brands and produces revenue from sales for sellers while proffering a solution to a pain point in the market.

Two major factors that have influenced Brimore's growth and success in Egypt's e-commerce market are its financial investments and collaborative network.

1. Seed Round and Pre-Series A Round: In an April 2019 seed round that was co-led by Algebra Ventures and Endure Capital, with about 500 Start-ups participating as well as Flat6Labs and angel investors, Brimore got up to $800,000. Later, the company got $3.5 million in a May 2020 pre-Series A round, with Algebra Ventures leading the round again with DisrupTech, Vision Ventures, and existing investors 500 startups and Flat6Labs. These prior investments sourced the needed capital for Brimore to grow its platform, expand its team, and upscale.

2. Investor Support: Algebra Ventures, which was one of Brimore's early investors, has continued to support Brimore throughout its journey. Algebra Ventures has provided Brimore with monetary backing, guidance, and industry expertise. This has been quite beneficial to the growth and expansion of the company. Egyptian fintech - Fawry's acquisition of a minority stake in Brimore further strengthened the company's position in the market. This collaboration allowed Brimore to use Fawry's large network of business owners and access its digital payments and financial services, thus improving its distribution abilities and offers.

3. Potential Equity Investment: The International Finance Corporation showed interest in making an equity investment of about $5 million in Brimore. This is primarily to support the company's expansion efforts. This potential investment would provide Brimore with additional capital to improve its existing infrastructure, improve operational excellence, and increase its operations in Egypt. With the IFC backing, Brimore can accelerate its growth trajectory and strengthen its position as a top player in Egypt's e-commerce scene.

4. Infrastructure and Expansion: The company plans to allocate the latest funding towards building its infrastructure, improving its operational excellence, and expanding its presence in the country. By investing in infrastructural development and operational efficiency, Brimore aims to improve the performance of its platforms, streamline its processes, and deliver a smooth experience to its users. The company's expansion plans reflect its commitment to capture a larger market share and serve more customers across Egypt.

Business Model

Brimore's business model surrounds social e-commerce and the empowerment of small businesses and individuals to sell products via social media while providing end-to-end support. The principal components of Brimore's business model are:

1. Social E-commerce Platform: Brimore has a social e-commerce platform that helps connect suppliers with community sellers. It serves as a marketplace where suppliers list their products, and community sellers advertise and sell these products through social media channels like WhatsApp, Facebook, and Instagram.

2. Supplier-Seller Network: The platform facilitates connections between new businesses and micro-sellers who use their social networks to promote and sell products. This network-driven approach helps suppliers reach a wider audience and generate demand for their products while providing revenue-making opportunities for micro-sellers.

3. Order Consolidation: When community sellers get orders from consumers, they place consolidated orders through Brimore's platform. Brimore then manages the fulfillment process, which includes order processing, warehousing, and last-mile delivery. Milezmore, the company's end-to-end fulfillment business, handles warehouse operations and delivery logistics.

4. Revenue Generation: Brimore generates income through several schemes, which include commission fees, fulfillment services, and value-added services.

- Commission Fees: Brimore may charge suppliers a commission fee for each sale done through its platform.

- Fulfilment of Orders: The company gets revenue from its end-to-end fulfillment services, charging fees for processing orders, warehousing, and delivery.

- Value-added Services: Brimore also earns funds by offering value-added services like marketing support, analytics, and training to suppliers and sellers for a fee.

5. Profit Margin for Sellers: The platform ensures community sellers earn a profit margin of 25% to 40% on sales. This incentive to actively publicize and sell products provides sellers with a source of revenue.

6. Diversification and Expansion: Brimore diversifies its revenue streams by expanding into other areas, like fulfillment services through Milezmore. The company also wants to explore

opportunities for expansion in the region, especially into markets in Africa and the Middle East, for future growth.

Features and Expansion

The features distinguishing Brimore as a top social e-commerce platform in Africa are;

1. Supplier-Seller Network: Brimore supports connections between new brands and sellers via its digital platform. This approach will help suppliers reach a wider audience and empower sellers to sell their products within their social circles.
2. Social Media Integration: The platform quickly joins popular social media channels like WhatsApp, Facebook, and Instagram. Through this means, micro-sellers can share promotional material and engage with their networks. This enhances visibility and promotes product discovery among potential customers.
3. End-to-end Fulfillment: Brimore grants end-to-end fulfillment services to its customers. These services consist of order processing, warehousing, and last-mile delivery through its subsidiary, Milezmore. This reduces the logistics procedure and ensures timely delivery to the customer's doorstep.
4. Commission-based Revenue Model: The platform has a commission-based revenue model, where it charges suppliers a commission fee for each sale made via its platform. This incentive for suppliers to use Brimore's network gives the company a sustainable revenue stream.
5. Profit Margin for Sellers: Brimore ensures all micro-sellers using its platform earn a substantial profit margin ranging from 25% to 40% on sales. This makes sellers actively publicize and

sell products, pushing sales growth and advancing long-term partnerships.

6. Diverse Product Categories: Brimore offers a variety of products across categories, such as household goods, personal care, fashion, electronics, furniture, and food and beverages. This diverse product choice range helps people with varying preferences and needs.

7. Empowerment of Women Sellers: While not explicitly targeted, Brimore has observed that most of its sellers are women. The platform provides women with income opportunities and economic empowerment, thus contributing to gender equality and community development.

8. Regional Expansion Plans: Brimore's future plot includes plans for regional expansion within Africa and the Middle East. This expansion plot aims to tap into underserved markets and capitalize on the growing e-commerce sector in the region.

Case Study - RETAILO: An App For Retailers to Restock their Shops

Retailo - e-Commerce Platform for Small and Medium-Sized Retail Business

Retailo, a Saudi Arabia-based B2B e-commerce company, has recently raised $15 million in funding from a group of new investors, including Yusuf Bin Kanoo Group, Technology Group, and Majd Digital, along with returning investors such as Aujan Group Holdings, Shorooq Partners, Abercross Holdings, Graphene Ventures, and others.

It was founded in 2020 by Talha Ansari, Wahaj Ahmed, and Mohammad Nowkhaiz. It operates as a next-day delivery platform for retailers and restaurants with a catalogue of 5,000 stock-keeping units.

The latest funding round will strengthen Retailo's existing operations, expand into new areas, and deepen its presence in Saudi Arabia. Retailo has also secured partnerships, such as one with a technology distribution partner—Dtonic, a South Korean data solution company. Retailo plans to use the fresh funding to enlarge its supplier network and customer base as it continues to grow in the Saudi market.

The Saudi Arabian retail industry is projected to be worth about $100 billion, and it provides for a populace of more than 35 million consumers. The company intends to digitize and completely transform Saudi Arabia's retail supply chain by using its digital distribution platform to connect suppliers with retailers/restaurants across the region. The company has significant traction from global brands who are looking to enter the Saudi Arabian market and distribute seamlessly.

Talha Ansari, the founder and CEO of Retailo, stressed the company's commitment to creating a stress-less digital distribution network for businesses, thereby contributing to the Kingdom's Vision 2030 initiatives. The company's advanced technology has made it one of the key players capable of shaping the future of the Saudi retail economy.

Yousef Albabtain, principal and Saudi country manager at Shorooq Partners, praised Retailo's dynamic leadership and agility in transforming the traditional retail sector in Saudi Arabia. Shorooq Partners is deeply committed to supporting Retailo as the leading tech-enabled B2B distribution company in Saudi Arabia.

Business Model

The business model of the Retailo start-up is basically about connecting suppliers with retailers and restaurants in Saudi Arabia via its virtual platform. The following details an outline of the company's business model:

1. Virtual Platform: Retailo has a centralized hub as a market that connects suppliers and wholesalers with retailers/restaurants for business.
2. Supplier Network: A vital part of Retailo's model is the network of suppliers and wholesalers that includes local, regional, and global brands. These suppliers have several products in different categories, allowing retailers/restaurants to access a vast inventory.
3. Retailer/Restaurant Networks: Retailo forms collaborations with retailers/restaurants across Saudi Arabia. These partnerships enable Retailo to onboard businesses on its platform, thus increasing its customer base and opening a way for higher demand for suppliers' products.
4. Next-Day Delivery Services: Retailo offers next-day delivery services for retailers and restaurants. By leveraging its logistics,

Retailo ensures timely order fulfillment, increasing customer convenience.

5. Credit Facilities: The company offers its customers credit facilities that allow retailers and restaurants to make purchases on credit terms. This helps businesses manage their cash flow and inventory without immediate financial reward.

6. Revenue Model: Retailo generates revenue through various channels. These include;

- Commission Fees: Suppliers and wholesalers are charged a commission fee for transactions on the Retailo platform. This fee is based on the transaction's value.

- Subscription or Membership Fees: The company offers suppliers premium subscriptions or memberships. These give them access to more benefits for a recurring fee.

- Value-Added Services: The company also provides value-added services like data analytics, marketing support, and inventory management to suppliers and retailers at a cost.

7. Technology Partnerships: the company enters into partnerships with tech partners to improve its platform offers and provide more solutions to its customers. These collaborations help the company leverage technologies like AI and data analytics to improve its operations and customer experiences.

Features

As a top business-to-business digital company in Saudi Arabia, Retailo offers several benefits that distinguish it in the market:

1. Next-Day Delivery: Retailo has next-day delivery services for retailers and restaurants. This is done to ensure timely order fulfillment and efficient supply chain management.

2. Product Catalogue: The company has a catalog of over 5,000 stock-keeping units (SKUs) and offers a wide range of products across various categories to cater to the needs of retailers and restaurants.

3. Supplier Network: Retailo has an enriching network of suppliers and wholesalers. This means that retailers and restaurants have access to a wealth of products from local, regional, and global brands.

4. Digital Platform: The platform is a digital market where suppliers can distribute their products to retailers/restaurants across Saudi Arabia. It is a simple tech solution for suppliers to reach a greater audience.

5. Technology Partnerships: Retailo collaborates with tech partners like Dtonic, a South Korean data solution company, to improve its platform's capabilities and provide advanced retail technology solutions to its customers.

6. Credit Facilities: Retailo offers credit facilities to its customers, which allows retailers and restaurants to make purchases on credit and properly manage their cash flow.

7. Data Analytics: The platform provides data insights to help retailers and restaurants make informed buying decisions, identify trends, and optimize inventory management strategies.

8. Expansion: Retailo is currently dwelling on expanding its operations within Saudi Arabia and deepening its presence in the market. The company aims to enlarge its customer and supplier network, and further solidify its position as a leader in B2B digital distribution.

Partnerships

Retailo has quite a number of partners as it continues to collaborate with various entities to improve its offers, expand its buying audience, and strengthen its position in the market. Some of such collaborations are;

1. Yusuf Bin Kanoo Group: Retailo recently secured investment from the Yusuf Bin Kanoo Group, a prominent Bahrain-based conglomerate. This partnership not only provides financial backing but also leads to greater expansion opportunities.

2. Technology Group and Majd Digital: Retailo received investment from Technology Group and Majd Digital, showing the confidence of both firms in Retailo's business model and growth potential. These collaborations bring more expertise and resources to Retailo.

3. Aujan Group Holdings: This is an existing investor who continues to support Retailo by aligning its interests with the company repeatedly.

4. Shorooq Partners, Abercross Holdings, and Graphene Ventures: These are returning investors who continually reaffirm their commitment to Retailo's vision and contribute to its growth journey via financial input and stability.

5. Dtonic: Retailo is in a technology distribution partnership with Dtonic, a leading South Korean data solution company that helped the former use Dtonic's advanced retail technology to improve the capabilities of the platform and offer value-added services to its customers.

6. Retail Partners: Retailo has a network of retailers/restaurants across Saudi Arabia, which helps to boost demand for Retailo's platform and services and provides retailers and restaurants with access to a host of products and next-day delivery services.

7. International Brands: Retailo partners with and offers products from several top global brands on its platform. Such partnerships increase Retailo's product catalogs and attract customers who are seeking quality products from well-known brands.

Case Study - FLOWARD: An App to Send Flowers and Gifts

Floward - Online Flowers and Gifts Delivery

Floward is an online flower and gift delivery shop in Kuwait that offers a convenient way for customers to buy and send flowers, gifts, and other related products. The co-founder, chairman, and CEO of Floward – Abdulaziz Al Loughani, went on a mission to transform the gift industry in the Middle East and North Africa. That expedition resulted in the establishment of Floward six years ago, and since then, the company has metamorphosed into a prominent online flower and gift company with a heavy presence across nine countries, including the GCC, Egypt, Jordan, and the U.K.

Features

1. Digital Representation of Tradition: Floward solves the lack of proper digital representation of customs, traditions, and cultural aspects of gifting in the Arab world. It's a convenient online platform where customers can shop for flowers and gifts and preserve the values of generosity and kindness ingrained in Arab culture.
2. Product Range: Floward offers several products, including fresh-cut flowers, cakes, chocolates, accessories, and perfumes. The wide product range was designed to accommodate various occasions so customers can find the perfect gift for their loved ones.
3. Operational Efficiency: Floward's operations are complex but efficient. The company buys flowers directly from farmers and growers worldwide, primarily from South America, East Africa, and Europe. It operates 21 fulfillment centers and employs over a thousand people to ensure timely delivery.
4. Sales Growth: Despite operational challenges, Floward has experienced exponential growth in sales, which have multiplied

nearly 50 times in three years. The company is on the way to achieving more profitability, thus demonstrating its potential in the market.

5. IPO Plans: Floward has plans to go public within 18 to 24 months, with a pre-IPO Series C funding round of $156 million. The company is planning for its IPO, and its plans include establishing governance requirements and appointing financial and legal advisors.

6. Strategic Expansion: Floward is focused on solidifying its position in current markets instead of expanding into new locations. It also partakes in mergers and acquisitions to promote its offers and stand out from its competitors.

7. Experienced Leadership: Abdulaziz Al Loughani, the co-founder of Floward, is the company's leader with a wealth of experience in entrepreneurship and investment. His previous successes, Talabat and Faith Capital, provide valuable insights and direction for Floward's growth and development.

Business Model & Expansion

Floward operates on a business model that dwells on the provision of convenient and innovative online flower and gift delivery services in the MENA region. Here's Floward's business model:

1. E-Commerce Platform: Floward is an e-commerce platform accessible through its website and mobile app. The platform is a market where customers can search, pick, and buy flowers, bouquets, and gifts for different events.

2. Product Offering: Floward offers gifts, including freshly cut flowers, cakes, chocolates, accessories, and perfumes for birthdays, anniversaries, weddings, graduations, and more.

Thus, customers have more than enough choices to suit their gifting needs.

3. Supply Chain Management: Floward tries to ensure the availability and freshness of its product offered by a complicated system. The company gets flowers directly from farmers and growers worldwide, primarily from South America, East Africa, and Europe. It operates 21 fulfillment centers to handle orders and ensure timely delivery.

4. Order Fulfilment and Delivery: Upon receipt of orders, the platform arranges and delivers the products using its specialized refrigerated fleet. The company offers same-day delivery services, so customers can receive their orders on time. The company's delivery network ensures that orders are delivered to customers with care.

5. Customer Experience: Floward prioritizes customer satisfaction by providing a stress-free, user-friendly online shopping experience. The platform is designed to make browsing, picking, and buying products convenient for customers. Floward also has a dedicated customer support team to answer questions, offer assistance, and provide a smooth market experience.

6. Revenue Generation: The company generates revenue primarily from the sale of flowers, bouquets, and gifts to customers. However, they also charge delivery fees based on factors like location and timing. Premium services or customized options are also available for which customers can pay an additional fee.

7. Expansion: Floward wants to expand its presence in the market and grow its customer base in the Middle East. The company plans to strengthen its position in existing markets by improving its product offers and operational efficiency as well as investing in marketing and other promotional activities.

Floward's model is to use efficient supply chain management and a customer-centered approach to provide high-quality, convenient flower and gift delivery services to customers in the Middle East.

Other Notable eCommerce Players

Eyewa

Millennial Brands

Nana

Capital

CHAPTER 3

Healthtech Startups Pioneering Healthtech Innovation

The emergence of e-health solutions is changing the health sector for the better in the GCC region. Digital technology has now given patients the power to actively engage in their care and manage chronic conditions, creating a profusion of opportunities for healthcare practitioners and investors in the region. Several e-health tactics, such as virtual consultations, telemedicine, remote monitoring services, and online diagnostics, have been adopted, all part of the innovative wave of health technology.

There is substantial potential for leveraging digital health solutions to improve patient outcomes in both Saudi Arabia and the UAE, especially in areas like chronic disease management, diagnostics, and preventive care. A recent Big 3 consulting firm forecast predicts that the combined digital health market in these countries could amount to US$4 billion by 2026. As much as this shows significant growth prospects, the UAE has already seen the set up of 257 HealthTech start-ups. These statistics show that the Middle East is a vibrant environment for innovative healthcare.

Several state initiatives, like the Public Investment Fund's investment in Babylon Health, a telemedicine company that provides AI-powered virtual consultations and chronic condition management, go ahead and show the wide acceptance of digital health in the region. The high demand for healthcare services is driven by population growth and aging demographics, and it continues to fuel investments from both the public and private sectors, reflecting a fertile ground for start-ups and investors.

The MENA healthcare market is experiencing remarkable growth, driven by the rising demand for healthcare services and excessive healthcare expenditures. According to statistical projections, the MENA health market will likely ascend at a compounded annual growth rate (CAGR) of 11.7% and reach $243.6 billion by 2023. In the Gulf Cooperation Council (GCC) countries, healthcare expenditures reached $105 billion in 2022, and the UAE alone accounts for more than half of this spending.

The UAE, to be specific, has emerged as an outstanding player in the region's health system, and even the government is the State's primary health investor. In recent times, there's been a noticeable increase in private sector participation in healthcare, which shows growing interest and rising investment in the sector.

The healthcare system in the Middle East is quickly changing, and there are myriad ways to approach health management and differing tiers of access to healthcare and health expenditure in different countries.

Despite these differences, several Middle Eastern countries, including Israel, the UAE, Oman, and Turkey, have been recognized for their efficient healthcare, placing them among the top 20 in Bloomberg's Health-Efficiency Index.

The healthcare sector in the Arab region has wholly aligned with international trends and, as expected, has grown exponentially, especially since the COVID-19 pandemic.

Healthtech start-up valuations have seen a significant increase. Several healthtech organizations in the region achieved a cojoined value of more than $1.5 billion in 2022, which is about twenty-two times higher than the 2016 year. This, in effect, shows the exacerbating importance of technology in solving health issues and improving health outcomes.

Only a handful of healthtech start-ups established in the 1990s and 2000s have increased value; meanwhile, the start-ups established after 2010 are the most highly valued nowadays.

This proves that non-negligible improvement in technological innovation occurred in the past decade, and it aptly shows the transformative potential of health in the Middle East.

The Arab region's healthcare market offers many opportunities for investment and innovation. This is due to the upward surge in demand for healthcare services, rising healthcare expenditures, and the overwhelming acceptance of the importance of technology in improving healthcare delivery.

Rise of HealthTech Start-ups

The Arab world consists of countries in the Middle East and North Africa. This region is home to several diverse populations with varying healthcare needs.

The importance of innovation in healthcare delivery is seen in situations like rapid urbanization, an aging population, and an exacerbating prevalence of chronic illnesses. Healthcare start-ups have sprung up as top players in providing solutions to healthcare issues by using technology to improve people's access, affordability, and quality of healthcare services.

Key Areas of Innovation

Several health tech organizations in the MENA region are promoting tech innovation in different areas of healthcare, which include:

1. Remote Care: The use of telemedicine has increased tremendously, as it helps patients consult with healthcare providers remotely without the need for physical meetings. Presently, health tech start-ups are developing platforms that provide virtual consultation, remote monitoring solutions, and telemedicine-enabled diagnostics. This expands access to healthcare services in underserved areas.
2. Health Information Systems: It is expedient to properly manage healthcare data to improve the care of patients and treatment outcomes. HealthTech start-ups are developing electronic health record systems, health information exchanges, and other health management platforms to narrow data collection and reduce storage and sharing among healthcare providers, thus improving care coordination and decision-making.

3. Digital Health Monitoring: This area encompasses the use of wearable devices and health monitoring applications, which enable individuals to take control of their health and well-being. This is why some Health Tech start-ups have developed wearable sensors, health apps, and devices that help users track signs, monitor severe conditions, and receive personalized health insights that promote preventive care and early intervention.

4. Health Analytics: In this area, data analytics and AI are being used to completely transform healthcare delivery via insights from large amounts of health data. Some Health Tech start-ups now use AI algorithms and machine learning models to analyze medical images, suggest treatment outcomes, and pick the best treatment plans. These are all aimed at improving precision in medicine and personalized care.

Challenges and Opportunities

While the Health Tech start-up space in the Arab world holds immense promise, it also faces several challenges. Some of the major issues that constitute quite a hassle in navigating the health technology space are legal impediments, data privacy doubts, and interoperability issues. Others include low finances, a dearth of talent shortages, and market fragmentation. All these are serious hindrances to scalability and growth in the healthcare space.

No matter the challenges, health tech start-ups in the Arab world are ready for momentous growth and expansion. The enormous population, exacerbating healthcare costs, and the State initiatives pushing e-health altogether provide a wealth of opportunities for innovation and investment in the health technology space.

The HealthTech start-up scene in the MENA region is undergoing extemporaneous growth and innovation, which, in turn, is being fuelled by the confluence of health, entrepreneurship, and technology. While focusing on providing solutions to health issues and improving the rate of positive results and healthcare delivery generally, we can boldly say that health tech start-ups are ready to play a transformative role in shaping the region's future of healthcare.

The fact remains that the health tech space in the Arab world is still evolving, and one factor that can push it upward even further is collaboration amongst start-ups, healthcare practitioners, policymakers, and investors. The truth is that this singular act will be holy in unlocking the full potential of HealthTech innovation in the Arab region.

Case Study - VEZEETA: The Champion of Healthtech Innovation

Vezeeta - Revolutionizing Healthcare Access in the Middle East and Africa

One of the most impactful changes in Arab's healthcare industry has been the establishment and upward trajectory of Vezeeta, a top e-health platform founded by Amir Barsoum, who launched the company with the vision of transforming the healthcare booking method in the MEA region. While setting up the company, he planned to make healthcare more accessible and efficient for more people while narrowing operation options for healthcare practitioners. This start-up has completely changed how patients access healthcare services in the region. This initiative has become one of the top e-health platforms in the region, and it has impacted how healthcare services are accessed and delivered.

The Genesis of Vezeeta

Vezeeta's exodus started in Egypt in 2012, immediately after it was established by Amir Barsoum. The platform was birthed out of a desire to close the gap between patients and health practitioners by offering a solution to the issue of accessing health services. Vezeeta began as an online booking space for medical appointments but quickly metamorphosed into a digital healthcare system. This change was prompted by the pressing need for more accessible, efficient, and patient-centric healthcare services in the MEA region.

Services and Solutions

The Vezeeta platform has many services that attend to the different needs of patients and healthcare practitioners. For patients, it is an easy-to-use interface to search for doctors by specialty, location, and insurance coverage, book appointments online, and access telemedicine. This connotes its usefulness in reducing waiting times and enhancing the healthcare experience for patients.

For healthcare personnel, Vezeeta is the home of software solutions to streamline operations, manage patient appointments, and optimize their online presence. These tools improve the efficiency of healthcare delivery and also enhance the quality of care provided to patients.

Impact and Achievements

So far, Vezeeta has had a tremendous impact on the healthcare industry in the MEA region. The platform has successfully brought thousands of doctors across various specialties onboard, thus assisting millions of patients in accessing healthcare services more efficiently. The use of the platform has gone beyond Egypt to Saudi Arabia, Jordan, Lebanon, and so on. In these locations, the platform is still uniquely improving healthcare access.

One of the top achievements of the Vezeeta company is its contribution to improving accessibility to good healthcare. By making the appointment booking process digital, Vezeeta has increased access to healthcare services, especially in underserved regions. The platform's e-health services were also really helpful during the Coronavirus pandemic as they facilitated medical consultations for patients from the comfort of their homes.

Challenges

Although it has been highly successful, Vezeeta's exodus has had its fair share of challenges. Initially, both patients and healthcare practitioners who were used to traditional healthcare methods and access opposed e-health solutions in the MEA region. The company overcame these challenges through its extensive awareness campaigns and enlightenment programs and by showing the enormous benefits of digital healthcare.

Another issue was the extant legal regulations that did not allow digital healthcare in most MEA countries. Vezeeta had to work with healthcare authorities and regulators to find a way around these challenges by advocating for laws that support the use of telehealth services.

The Vezeeta platform is still innovating and introducing more advanced healthcare solutions, including diagnostic AI-driven machinery and personalized healthcare recommendations. Their goal is to become the topmost eHealth platform not only in the MEA region but all over the world by continuing to break down hindrances to healthcare access and improving patients' results.

Vezeeta also plans to deepen its impact by spreading its tentacles to more countries within the MEA region and beyond. The company is also trying to strengthen its collaborative relationships with healthcare personnel, insurance companies, and governments to create a more integrated healthcare system that benefits all stakeholders. Vezeeta is leading the e-healthcare revolution in the MEA. It has improved access to healthcare, efficiency, and quality for millions of patients and healthcare practitioners. As the company continues to expand, its impact on the healthcare industry is expected to grow.

Business Model

Vezeeta has carved out a niche for itself in the online healthcare space via its novel business model, which uses cloud-based technologies to offer accessible healthcare solutions across the Middle East and Africa. As stated, the model was created to reduce patient-medical practitioner interaction and facilitate a more effective healthcare experience.

Vezeeta's cloud-based platform is the core of its business model, enabling self-subscribing and automated, remote, and virtual onboarding for users. This method simplifies the process for healthcare practitioners to join the platform and ensures that patients have easy access to several healthcare services from the comfort of their homes. The cloud-based nature of Vezeeta's solutions connotes that the platform can easily scale across different countries and adapt to local healthcare needs and regulations without the need for physical infrastructure.

Vezeeta operates in four different countries, which reflects the adaptability and scalability of its business model. This international presence also shows the company's ability to navigate different healthcare systems and consumer needs so it can tailor its offerings to meet the specific demands of each market.

The company's revenue model is multifaceted, combining transaction fees with subscription services. It charges a transaction fee for bookings made through the platform, including health practitioner consultation, services, surgeries, and simple diagnoses. This fee structure is in line with the company's interests and those of its users since Vezeeta benefits directly from facilitating more healthcare interactions.

As an addition to transaction fees, Vezeeta also offers Software as a Service (SaaS) solutions to medical care providers. These solutions cover a mass of software tools designed to improve the operational efficiency of healthcare institutions, including appointment scheduling, patient management, and digital consultation. Medical practitioners pay a monthly subscription fee for access to these SaaS solutions, and this ensures a steady stream of recurring income for the company. Again, the company's business model is proof of the potential of digital technology to transform the health industry.

Features

Vezeeta's platform comes with several characteristics and benefits that are answers to the myriads of traditional challenges that medical practitioners and patients face in the Middle East and Africa. These characteristics are designed to reduce the healthcare process and make it more efficient, accessible, and user-friendly for all parties. Some of these characteristics are:

1. Virtual Appointment Booking: Patients can now search online for doctors of their choice by specialty, location, insurance coverage, and so on, then book appointments online at their convenience without the need for endless phone calls and waiting times.

2. E-health Services: Medical practitioners can now offer virtual consultations so people can receive medical advice without physical visits to any healthcare facility. This is beneficial during pandemics/epidemics and infectious diseases, as well as for those with mobility issues and those in underserved areas.

3. Medical Practitioner Software Solutions: Vezeeta has a suite of SaaS tools that help medical practitioners and their institutions manage bookings, patient records, and online consultations. These tools are made to improve operational medical efficiency and patient care.

4. User Ratings and Reviews: Patients can rate their experience and leave reviews for medical service providers. This helps others make informed decisions based on prior feedback.

5. Insurance Integration: The platform supports various insurance plans, making it easier for patients to find health practitioners within their network.

6. Multi-Country Scalability: The app is designed to operate in different countries, so it adapts to local healthcare systems and regulations.

The attendant benefits are:

Greater Accessibility: Vezeeta's platform makes it easy for people in remote and underserved areas to access healthcare through online bookings and telehealth consultations.

Greater Efficiency: The platform also reduces the wait time for appointments and the 'bureaucracy' associated with healthcare, thus making the healthcare process more efficient for both patients and health practitioners.

Greater Transparency: With its user ratings and reviews, Vezeeta promotes greater transparency in the healthcare sector, allowing patients to make informed decisions based on the experiences of others.

Better Healthcare Management: Health practitioners benefit greatly from Vezeeta's software solutions because they help manage patient appointments, record-keeping, and consultation better, which ultimately translates to improved patient care.

Convenience: The platform offers ease in managing healthcare needs virtually from the comfort of one's home.

Multiple Insurance Support: By granting patients access to several insurance plans, Vezeeta makes it easier for them to find compatible healthcare providers.

In conclusion, Vezeeta is addressing gaps in the MEA healthcare system with a more patient-centric approach that aligns with the needs of modern healthcare consumers and practitioners.

Expansion

While we admit that Vezeeta's innovation is top-notch, one pillar that has been instrumental in the growth of the start-up is the partnerships it has entered into. To say the least, these collaborations have been extremely important in improving Vezeeta's services, expanding its reach, and increasing the quality of healthcare services offered to patients. Here are some examples of the partnerships Vezeeta has pursued and the impact they have had on its success;

Medical Practitioners and Health Institutions

The foundational partnerships Vezeeta has had are those with medical practitioners and healthcare institutions. The company now boasts a vast network of doctors, physiotherapists, specialists, and healthcare facilities, ensuring its users have access to a wide range of medical services. This kind of partnership helps healthcare service providers manage bookings, keep patient records, and facilitate online consultations through Vezeeta's platform, thus resulting in more efficiency and greater patient care.

Insurance Companies

Vezeeta has collaborated with some insurance firms to streamline the process for patients to find and book appointments with medical professionals within their insurance network. These partnerships benefit patients by making it easier for them to understand and utilize their insurance coverage. They also benefit medical service providers by facilitating the billing and reimbursement process.

Technology and Telecommunications Firms

Vezeeta has also partnered with other technology and telecommunication firms. This has proved to be beneficial to the company in terms of technological infrastructure and expansion of its services. These collaborations have helped the company develop and implement advanced features like AI-driven diagnostics, e-health service offers, and secure, cloud-based patient management systems.

Government and Regulatory Bodies

Engaging with government and regulatory bodies has also been of great help to Vezeeta, as it has helped the company understand the inroads into each country despite the complicated labyrinths of health laws and regulations in different countries. Through such government collaborations, Vezeeta has aligned its operations with local healthcare policies and standards. This has, in turn, facilitated its expansion into new markets. Collaborations with public health programs have also enabled the company to contribute to national health goals like improving public access to quality healthcare and supporting disease surveillance and control.

International Health Organizations

Through its collaborations with international health organizations, Vezeeta now has the opportunity to contribute to and benefit from global health programs. These partnerships have helped Vezeeta gain insights into eHealth best practices, participate in international health campaigns, and access funding and support for expanding into underserved regions.

Impact of Partnerships

As stated previously, Vezeeta has been involved in collaborations that have been instrumental in its mission to transform the healthcare access and delivery system in the MEA region. Through numerous collaborations, Vezeeta has created a healthcare system that benefits all parties involved—from patients and healthcare practitioners to insurance companies and the government. These partnerships improve Vezeeta's services and also contribute to the broader aim of improving healthcare outcomes and accessibility for the vast populace in the region.

Case Study - ALTIBBI: Empowering Healthcare through Innovative Digital Solutions

Altibbi - Telemedical Consultation Services

Altibbi is an e-health platform in the Middle East and North Africa. The company offers medical information and healthcare services in Arabic.

It was established by Jalil Allabadi in Amman, Jordan, in 2011. The inspiration for Altibbi emanated from a desire to meet the need for reliable and accessible medical information for people who speak Arabic. The idea was originally from a medical dictionary crafted by Jalil's father, Dr. Abdulaziz Allabadi, upon returning from Germany in the year 2004. They recognized the demand for medical information in Arabic, so they expanded this concept into a more extensive platform.

AlTibbi's services include medical articles, a glossary, a Q&A section, the latest medical news, telehealth services, and consultations. The company has been involved in several partnership relationships all to improve its offers and audience. These relationships include working with Mawdoo3, the top online Arabic content provider, to increase the scope of content that is usually made available to people in the region.

The company has also worked with the Ministry of Health and the Ministry of Communications and Information Technology to provide 24/7 e-medical consultations in Egypt. AlTibbi has partnered with the United Nations Development Program in Egypt and the Egyptian Telecommunications Company for a campaign to offer healthcare consultations to patients in remote areas.

The company also worked with the Royal Health Awareness Society to provide special care to people with chronic diseases and also entered into relationships with companies like Reckitt to promote awareness of health issues.

AlTibbi's move from a simple start-up to a top e-health platform in the Arab world involves several funding milestones. Initially, it was self-funded by the founder, and contributions were made from friends and family. It moved from that to its first major investments in 2015 from Middle East Venture Partners (MEVP) and Dash Ventures.

This financial boost helped to improve interactions between doctors and patients and expand the user base across five Arab countries. Further investment rounds in 2017 and a financial boost of $44 million in 2023 supported AlTibbi's development of comprehensive healthcare solutions and expansion into new markets.

 Innovations like the Al-Tibbi application, introduced in early 2021, have made medical consultations more accessible and convenient. The app features advanced technologies like AI for transcribing audio recommendations and reading vital signs through the mobile phone camera.

AlTibbi also launched the Tebi Mama app, which offers prenatal and postnatal care. The launch also included the Tebi Academy, which was created in collaboration with Primary Care International for doctor training.

The platform has more than 12,000 accredited medical doctors across various specializations and countries. AlTibbi has made significant strides in providing accessible healthcare information and services. It has registered tens of thousands of users for its phone consultation services, and it still attracts millions of monthly website visitors. The platform's commitment to offering quality medical content in Arabic was recognized with the first prize in the health category by the Arab E-Content Award in Bahrain in 2013.

Business Model

AlTibbi has a model that uses digital technology to help Arabic speakers in the MEA region access many healthcare services and information primarily directed at them. This is quite a multidimensional model as it involves several revenue streams and services designed to meet the needs of both healthcare service providers and patients. This is AlTibbi's model in detail:

Subscription Services: AlTibbi offers subscription-based services for individual users and medical professionals. Individual users subscribe to access premium content like in-depth medical articles, personal health guidance, and advanced e-health consultations. Medical professionals and institutions also subscribe to AlTibbi to manage appointments, electronic health records, and teleconsultations.

e-Health Consultations: An important component of AlTibbi's model is its e-health services, which allow patients to consult medical professionals via the Internet or phone. These services are provided on a fee-per-consultation basis or through subscription packages, making healthcare increasingly accessible to individuals in remote or underserved areas.

Advertising: AlTibbi also generates income via advertising and sponsored content. The company allows health-related businesses and pharmaceutical companies to advertise their products and services to its audience via banner ads, sponsored articles, and product placements in the app and/or site.

Collaborations: AlTibbi expands its user base through partnerships with health institutions, government, and tech companies. These partnerships often involve co-developed programs like public health awareness campaigns and probably additional income streams via joint ventures or service contracts.

Content: AlTibbi uses its extensive database of medical content and user interactions to generate insights and analytics that are quite valuable to research institutions, healthcare providers, and pharmaceutical companies. Selling access to anonymized data sets obtained from user behavior and health trends is another income source that aligns with data privacy protection rules.

E-commerce and Online Pharmacy: Though this is not stated in the description, platforms like AlTibbi explore e-commerce opportunities like online pharmacies or markets for health and wellness products. This allows users to buy medications, supplements, and health-related products directly from the platform, thus adding an extra income stream while providing ease to patient consumers.

Investment and Funding: AlTibbi has received several rounds of investment, which have contributed to its development and upscaling. Investment is not an income stream, but it is an important part of AlTibbi's model, and it helps the platform innovate, expand into new markets, and improve its offerings.

Impact: AlTibbi's model is crafted to achieve profitability and social impact by improving access to health information and services among Arabic speakers. This dual focus on financial sustainability and social value creation has placed AlTibbi as a top platform in the MENA region's virtual health scene.

Features & Benefits

As an e-health company with several advantages, AlTibbi is crafted to meet users' needs in the MEA region. Its characteristics are made to improve access to concrete medical information and healthcare services, thus opening the path for better healthcare experience for both patients and healthcare service providers. This is an overview of AlTibbi's characteristics and the merits they provide;

Features

- Virtual Medical Consultations: AlTibbi allows users to consult medical practitioners online. This makes for convenience and accessibility, especially for those living in remote areas or with limited physical mobility.
- Medical Articles: The company has a vast library filled with medical articles and information vetted by medical professionals to educate users on various health topics in Arabic.
- The Questions and Answers Section: This section allows users to ask health-related questions and receive answers from qualified medical professionals on the platform. This is important for social bonding, interaction, and support.
- Medical Glossary: The platform also has an all-encompassing medical glossary that helps users understand medical terminology in their native language.
- Latest Medical News: AlTibbi keeps its users updated on the latest news in medicine and health.
- Partnerships for Enhanced Content: AlTibbi collaborates with leading content providers like Mawdoo3 to improve the variety and depth of medical content available to users.

Benefits

Improved Access to Healthcare: AlTibbi makes healthcare more accessible to people, especially those living in areas with limited healthcare facilities, by offering e-medical services and consultations.

Educational Resources: the company's massive library of articles, glossary, and Q&A section provides valuable academic resources.

Informed Decision-Making: AlTibbi grants people access to reliable medical information and advice from professionals, which helps them make informed decisions about their health and care.

Convenience: Since AlTibbi is online, it offers its users access to a wealth of medical information online and e-medical consultations. These offers provide convenience to people who use the platform and save them time.

Community Support: the Q&A section and the ability to interact with medical professionals on AlTibbi give people a sense of community and support.

Language Accessibility: by providing its content in Arabic, AlTibbi solves the issue of the language barrier.

Expansion

AlTibbi's collaborations help to improve its service offerings and expand its reach within the e-health scene, particularly in the Middle East and North Africa. These partnerships span various sectors - government health departments, private healthcare institutions, technology companies, and educational/research institutions, all contributing to AlTibbi's mission. This is an overview of how partnerships contribute to AlTibbi's success:

Government Health Departments: getting into relationships with government health departments helps AlTibbi align its services with national health programs and regulations. This ensures that its offers can meet the needs and standards of each country it operates in. Such relationships often promote public health campaigns, awareness programs, and access to e-health services.

Private Health Practitioners: partnerships with private medical practitioners and institutions help improve the range of services available on AlTibbi's platform. Through these collaborations, AlTibbi can offer a more comprehensive directory of healthcare professionals for online consultations and bookings over a wide range of services. This benefits patients searching for convenient access to healthcare and helps health practitioners reach a larger patient base.

Tech Companies: AlTibbi makes use of the latest e-health technology and this is a result of its working relationship with tech companies that specialize in healthcare IT, telecommunication, and mobile app development. These partnerships help to develop and integrate innovative features like AI-driven diagnostic tools, secure patient data management systems, and advanced e-health services that help to boost the user experience.

Educational Institutions: Collaborations with educational/research institutions and medical associations are important to ensure the quality and reliability of medical information provided on AlTibbi's platform. These partnerships help curate and verify medical content, including articles, glossaries, and Q&A responses, to maintain high standards of accuracy. Furthermore, educational collaborations lead to developing training programs for healthcare professionals.

Content Providers: AlTibbi's partnerships with content providers like Mawdoo3 provide a diverse range of high-quality, Arabic-language medical content. These collaborations help ensure that users have access to a large repository of information that is both reliable and engaging for the patients that make up the Arabic-speaking population.

Impact of Partnerships: AlTibbi has gotten into partnerships that have been quite helpful in its ability to provide e-healthcare. Such partnerships support the platform's upscaling across the MENA region and also help improve the quality and accessibility of healthcare information and services for those who speak Arabic. Via its relationships with different stakeholders in the healthcare space, AlTibbi is well-placed to keep up with its goal of transforming access to healthcare and healthcare delivery in the region.

Case Study - AUMET: Digital Solutions for Pharmacy Management and Orders

Aumet - Solutions for Pharmacy Management and Orders

Aumet was founded by Yahya Aqel in 2015. The founder's vision for Aumet was to create a platform that would change the healthcare supply chain in the Middle East and North Africa region by connecting pharmacies with suppliers through a seamless B2B market.

Aqel's headship has shepherded Aumet to grow and enabled it to become a top player in digitalizing the healthcare procurement procedure and facilitating efficient, tech-driven connections between health practitioners and health product suppliers across the region.

Recently, Aumet, a health tech start-up based in Saudi Arabia, received a $7 million pre-Series A funding round. This was pretty amazing, and it shows the increasing interest and investment in the health tech sector in the MEA region.

Aumet has an AI-enabled B2B healthcare platform that gives bespoke tech solutions to medical service providers. This shows a lot of advancement in the digitalization of the healthcare supply chain.

The Aumet platform uses predictive analytics to forecast pharmacies' procurement needs. The company wants to gain cost savings and efficiencies in the supply chain. This act narrows down the procurement process and ensures that pharmacies are better equipped to meet the demand of the audience for healthcare products.

The participation of global venture capital and private equity firms AAIC, AIJ Holdings, and Shorooq Partners, as well as healthcare investors and pharmaceutical groups like the Cigalah Group, in this funding round shows confidence in Aumet's mission and its potential to positively impact the healthcare industry.

The funds from this round are marked for further development of Aumet's AI capabilities, expansion of its pharmacy network, and improvement of access to affordable healthcare products. This is in line with Aumet's broader vision of digitalizing the medicine and medical procurement cycle all over the world to secure the lives of billions of people.

Aumet's growth comes at a time when the MEA region is experiencing an increase in healthcare technology innovation. Similar start-ups like Egypt-based Yodawy and Vezeeta, as well as Cairo-based Grinta, have also secured funding. The latter reflects a robust and dynamic healthcare technology system in the region. These developments are tilting towards digital revitalization in healthcare, focusing on increasing accessibility, efficiency, and affordability of healthcare services and health products.

The success of these start-up companies and the inflow of investments into the health tech sector in the MEA region show the potential for innovative solutions to solve long-time problems in healthcare delivery and management. As these companies continue to grow, they will all play an important role in shaping the future of healthcare in the region and probably beyond using technology to improve patient outcomes and the efficacy of healthcare systems.

Aumet, the Jordan-born, and US-headquartered B2B healthcare marketplace acquired the Egypt-based healthcare supply chain solutions start-up – Platform. This was quite a milestone in Aumet's expansion strategy, particularly its entry into the Egyptian market. This strategic move not only broadened Aumet's geographical footprint but also strengthened its position in the region's healthcare space.

Platform One was founded in 2018 by Yossry Farghaly, Abdelhady Araby, and Osama Al-Atroush. From then on, it established itself as a key player in Egypt's healthcare supply chain, with over 600 pharmacies and 150 suppliers. Aumet, on the other hand, was established in 2015 by Yahya Aquel and has since connected thousands of pharmacies with suppliers across the MEA region. The acquisition deal details that Platform's team will spearhead Aumet's operations in Egypt by using their wide network and deep understanding of the local market to improve Aumet's service offers. This integration will yield the following benefits:

1. Wider Network: Aumet's acquisition of Platform expands its network - Platform's strong presence in Egypt combined with Aumet's regional influence is a merger that brings Aumet to over 5,000 active pharmacies across Saudi Arabia, United Arab Emirates, Egypt, and Jordan.

2. Greater Service Offerings: With Platform One's expertise, Aumet can enhance its product offerings in Egypt—Aumet Marketplace, Aumet Pharmacy POS and inventory management software, and Aumet Pay, a payment gateway for pharmacies and their suppliers. These offers are intended to disrupt traditional healthcare supply chains by automating and digitalizing processes for greater efficiency and reduced costs.

3. Strategic Market Entry: Egypt is a vital market location in the MEA region because of its large population and healthcare needs. The platform's established operations give Aumet a solid foundation for scaling its services in Egypt, in line with its goal of digitalizing the medicine and medical procurement cycle worldwide.

4. Synergy and Innovation: The acquisition fosters an innovative-ripe environment, as both Aumet and Platform One share a

commitment to tech-focused solutions for the health industry. This synergy will drive the development of new technologies and services that help solve the problems of the healthcare supply chain in the MEA region.

5. Leadership and Vision: Platform One's founders, Yossry Farghaly, Abdelhady Araby, and Osama Al-Atroush, bring their leadership to Aumet's operations in Egypt. This helps ensure that Platform One's vision and operational excellence will continue to influence Aumet's growth and service quality.

Business Model

Aumet has a model that uses artificial intelligence and digital technology to radicalize the healthcare supply chain with a sole focus on the business-to-business segment. This allows the company to cater to a wider field of stakeholders in the healthcare industry, including pharmacies, suppliers, manufacturers, and distributors. This is how Aumet's business model looks:

Core Components of Aumet's Business Model

1. B2B Marketplace: The core of Aumet's business model is its B2B market, which connects pharmacies with pharmaceutical suppliers. This platform enables the smooth exchange of products and information, enabling pharmacies to efficiently find and procure medical supplies.

2. AI-Enabled Predictive Analytics: Aumet uses AI to predict analytics for procurement processes. This helps pharmacies to forecast their product needs, manage their inventory more effectively, reduce stock-outs, and avoid overstocking. Predictive analytics also makes for significant cost savings and greater efficiencies in the supply chain.

3. Data Exchange: The platform is a data exchange hub where medical practitioners, manufacturers, and distributors can share and access information. This improves decision-making across the supply chain and service delivery.

4. Software Solutions: Aumet provides several software solutions to meet the needs of health service providers. These include Pharmacy Management Systems (PMS), point-of-sale (POS) systems, and inventory management software. All these are designed to streamline operations, improve cash flow, and automate ordering processes with features like 'Stock Auto-Replenishment.'

5. Aumet Pay: This online payment gateway makes for smooth transactions between pharmacies and their suppliers. Aumet Pay aims to disrupt traditional cash collection methods, offering a more secure, efficient, and transparent way to handle financial transactions within the healthcare supply chain.

Revenue Streams

Aumet's business model generates income through several channels:

- Subscription Fees: Pharmacies, suppliers, and manufacturers pay subscription fees to access Aumet's market and software solutions, thus benefiting from the platform's networking, analytics, and management tools.

- Transaction Fees: Aumet charges a specific fee for transactions done through its market, which includes selling pharmaceuticals and medical supplies.

- Software Licenses: The start-up also gets income by selling licenses for its software solutions to healthcare practitioners.

- Payment Processing Fees: The start-up earns fees for processing payments through Aumet Pay, which is a secure and efficient way for stakeholders to manage their financial transactions.

Advantages

Aumet's business model provides several advantages, which include:

- Efficiency and Cost Minimization: By digitalizing the procurement cycle and utilizing predictive analytics, Aumet helps pharmacies manage inventory more efficiently and reduces costs.
- Access to Markets: The platform helps suppliers and manufacturers gain access to a broader market, allowing them to reach more customers and expand their business.
- Data-driven Decisions: The company's access grants an expansive amount of data that enables all parties in the supply chain to make informed decisions and improve service delivery.

Features & Benefits

Aumet possesses several characteristics designed to narrow the healthcare supply chain, especially for pharmacies, suppliers, and manufacturers.

These features are built around improving efficiency, reducing costs, and increasing accessibility of healthcare products.

Here are the characteristics of Aumet's platform:

Features

B2B Marketplace: Aumet has a central market that connects pharmacies with pharmaceutical suppliers, enabling a stress-free transaction procedure. This market is designed to help with the easy discovery and procurement of medical supplies and pharmaceuticals.

AI-Enabled Predictions: The platform uses artificial intelligence to offer predictive analytics for pharmacies' procurement procedures. This helps predict product demand and allows pharmacies to optimize their inventory levels.

Data Exchange Platform: Aumet is a hub for information exchange between health institutions, manufacturers, and distributors. This supports brainstorming and decision-making by granting access to data across the supply chain.

Software Solutions: The platform offers software solutions such as Pharmacy Management Systems, point-of-sale systems, and inventory management software. These tools are designed to automate pharmacy operations.

Aumet Pay is an online payment gateway that facilitates transactions between pharmacies and suppliers. This aims to streamline the payment process and improve financial management within the supply chain.

Benefits

Streamlined Supply Chain: By connecting pharmacies directly with suppliers, Aumet narrows the healthcare supply chain, reducing the time and effort spent obtaining medical supplies.

Inventory Optimization: Aumet's predictions help pharmacies maintain optimal inventory levels and reduce the risk of stockouts or excess, reducing waste and costs.

Informed Decision-Making: Aumet's data exchange platform ensures that all parties in the supply chain have access to the information they need to make informed decisions.

Operational Efficiency: Aumet's software solutions automate operations like inventory management and ordering procedures in pharmacies, freeing up time and resources for other essential tasks.

Financial Management: With Aumet Pay, pharmacies and suppliers can manage their financial transactions more efficiently, thus benefiting from a secure, transparent, and streamlined payment procedure.

Expansion

Aumet's relationships are vital to its business model and growth. Its various relationships help the company to improve its offers and increase its presence in the market, especially in the healthcare and pharmaceutical space within the MEA region. These relationships range from link-ups with healthcare institutions and pharmaceutical companies to alliances with tech firms and financial institutions. This is how partnerships contribute to Aumet's success:

1. Healthcare Institutions and Pharmacies: Aumet partners with hospitals, clinics, and pharmacies to expand its user base and increase the transactions on its platform. These partnerships help the company inculcate its solutions into the systems of healthcare practitioners.
2. Pharmaceutical Companies and Suppliers: Collaborations with pharmaceutical manufacturers and distributors help the company offer several products in its market. These partnerships ensure pharmacies have access to a catalog of

medical supplies and pharmaceuticals, thus improving the efficiency of the supply chain.

3. Technology Firms: Aumet usually partners with tech companies, especially those that specialize in AI, data analytics, and cloud computing, to boost its tech infrastructure. Such relationships help further the development of advanced features like predictive analytics and AI-driven decision-making tools.

4. Financial Institutions and Payment Gateways: Aumet Pay helps streamline everyone's transaction procedure due to collaborations with financial institutions and payment gateways. These partnerships ensure secure and efficient payment processing, thus reducing reliance on traditional cash transactions.

5. Government and Regulatory Bodies: Aumet's relationships with government health departments and regulatory agencies help ensure its operations align with local regulations and standards. This kind of partnership also opens up opportunities for public health programs that further inculcate Aumet into the healthcare system of the MEA region.

Benefits of Partnerships

Market Access and Expansion: Partnerships help Aumet quickly enter new markets and upscale using the established presence and credibility of its partners.

Enhanced Product and Service Offers: Collaborations with diverse stakeholders help Aumet continue improving its platform and adding new features and services to meet changing user needs.

Operational Efficiency: By integrating with the systems of health institutions and health product suppliers, Aumet narrows down the procurement process to reduce costs and improve supply chain efficiency.

Innovation: Partnerships with technology firms drive innovation. This allows Aumet to implement the latest technologies in AI and data analytics. This is done to offer more sophisticated solutions to users.

Compliance: Working closely with government and regulatory bodies helps ensure that Aumet's operations comply with local laws.

Aumet's partnerships are the core of its mission to digitalize the medical procurement cycle worldwide. By collaborating with stakeholders across the healthcare, technology, and financial sectors, Aumet is properly positioned to radicalize the healthcare supply chain in the MEA region and beyond.

Case Study - DIAGNIO: Revolutionizing Women's Health with Cutting-Edge Technology

Diagnio - Transforming Women's Health Diagnostics

Diagnio is a health tech start-up platform that Marina Sol founded. The company is at the forefront of health-tech innovation, specifically targeting women's hormonal health and fertility.

This company offers express diagnostics solutions that can be utilized comfortably at home or in clinics. The importance of Diagnio's approach lies in its three foundational pillars:

1. Objective Data from Biosensors: Diagnio uses advanced biosensor technology to provide accurate, real-time data on women's hormonal health. The data then serves as the basis for all subsequent analyses and recommendations, ensuring that users receive insights grounded in reliable measurements.

2. Algorithms for Personalized Analytics: Based on biosensor data, Diagnio employs sophisticated algorithms to analyze and interpret the information. This process results in personalized analytics that consider each user's unique hormonal profile and health status, allowing for a deeper understanding of one's health and fertility status.

3. Actionable Advice from Online Consultations with Specialists: One of Diagnio's most impactful offers is its facilitation of online consultations with healthcare specialists. Users can receive actionable advice based on the analytics obtained from their biosensor data, allowing for informed health decisions and interventions. This feature democratizes access to specialist insights, making expert advice more accessible to women, especially those in physical areas with limited healthcare resources.

Diagnio's mission is to disrupt the traditional lab business model. This mission is fuelled by its use of technology to provide immediate, personalized health information. With a particular focus on the Gulf market, where infertility rates have hit a record high of 40%, Diagnio both proffers solutions to a critical health challenge and tries to close a gap in the market. The company's global ambitions are supported by a scalable model that could potentially transform how women's hormonal health and fertility issues are managed all over the world.

Diagno was established in 2022, and as such, it is a relatively new entrant in the health tech sector. However, the company's new method and focus on solving a pressing health issue makes it a promising player with the potential to make a substantial impact in women's healthcare, especially in fertility and hormonal health management.

The demand for health apps for women is increasing daily, with predictions showing the market will amount to USD 3.8 billion by 2023, an increase from the previous year.

The growing importance of digital health tools for women with the pandemic boosted the need for these apps by over 40%, stressing their role in women's health and wellness.

Diagnio enlisted an IT development team to create an app designed for effortless self-monitoring of women's health. The app offers the convenience of keeping track of health without regular doctor visits. The app represents a new era in women's health focused on fertility tracking, holistic health monitoring, and integration with ovulation tests to predict fertility windows accurately.

The development journey of Diagnio's women's health app was grounded in scientific research to improve lives through technology.

Insights from this trajectory include the importance of thorough market research, understanding health needs, and identification of the target audience—tech-savvy, health-conscious women in the reproductive age group in the UAE. Competitor analysis revealed that while the market is competitive, Diagnio stands out with its focus on integrating advanced diagnostic tools via a user-friendly interface. This approach fills a niche in personalized, home-based health monitoring.

Pre-development challenges included ensuring accurate health monitoring through a saliva analysis device, creating a user-friendly technology interface, ensuring data privacy and security, integrating with health systems, and navigating regulatory compliance.

The Diagnio app offers comprehensive women's health tracking and analysis, including menstrual cycle tracking, ovulation prediction, holistic health assessments, saliva-based fertility testing, and data integration with health platforms. These features aim to give users a complete picture of their health and fertility.

Post-launch analysis and user feedback highlighted the app's comprehensive approach and the need for more profound and informative questionnaires. Performance metrics and user feedback led to iterative improvements to enhance the app's functionality and user experience.

Developing Diagnio involved navigating technical challenges, such as data integration and ensuring accurate hormone level readings from saliva. The development process followed an agile methodology, incorporating flexible project management, efficient data handling, an intuitive user interface, rigorous quality assurance, secure data protection, and regulatory compliance.

Creating the Diagnio app underscores the significance of user-centric innovation in digital health technology. By focusing on user needs, incorporating feedback, and leveraging technology, Diagnio has successfully enhanced healthcare management for women. This journey offers valuable insights for those looking to venture into the digital health space, emphasizing the importance of specialized healthcare mobile app development and engaging medical professionals in the development process.

Business Model

Diagnio's business model provides innovative and accessible diagnostic solutions for women's hormonal health and fertility. This model is particularly tailored to meet the needs of the Gulf market, where there is a significant demand due to high infertility rates.

Diagnio integrates technology with healthcare expertise to offer a comprehensive service that includes objective data collection, personalized analytics, and actionable advice through online consultations. Here's a breakdown of its business model

Objective Data from Biosensors

Product Sales: Diagnio generates revenue by directly selling biosensors and diagnostic devices used by women at home or in clinics. These devices collect accurate and real-time data on various hormonal levels.

Personalized Analytics Algorithms

Subscription Services: Users subscribe to access Diagnio's platform, where personalized analytics are provided. This subscription model ensures a steady income stream while allowing users to benefit from continuous updates and improvements in the analytics algorithms.

Actionable Advice from Online Consultations

Consultation Fees: Diagnio monetizes the specialist consultations offered through its platform. Users seeking personalized advice based on their diagnostic results can pay for one-time consultations or multiple sessions.

Additional Revenue Streams

Partnerships with Healthcare Institutions: Diagnio can integrate its diagnostic solutions into its service offerings by partnering with clinics and other health institutions. The company can potentially earn revenue through licensing agreements or service fees.

Data Analytics Services: With the consent of its users, Diagnio could anonymize and aggregate the collected data to perform broader market or health studies. These insights could be valuable for pharmaceutical companies, research institutions, and healthcare policymakers, providing additional income.

Educational Content and Products: the platform could offer educational materials and courses related to women's health, fertility, and hormonal balance.

Market Focus

Gulf Market Orientation: Targeting the Gulf market, where there is a critical need due to high infertility rates, provides Diagnio with a focused approach that addresses specific regional challenges. This targeted strategy allows for tailored marketing efforts and product development to meet the unique needs of this demography.

Scalability

Global Expansion Potential: although it first focused on the Gulf market, Diagnio's business model and technological foundation allow for scalability. The universal nature of women's health concerns means the company has the potential to expand globally, adapting its offerings to different markets and regulatory environments.

Diagnio's business model is innovative in its integration of technology and healthcare, offering a personalized and accessible solution for women's health. By focusing on a market with a specific and urgent need, Diagnio positions itself for rapid growth and impact, with the flexibility to scale and diversify its income streams over time.

Features and Benefits

Diagnio's approach to women's hormonal health and fertility diagnostics is built on a foundation of cutting-edge technology and personal healthcare. The top characteristics and benefits of Diagnio's offs show its commitment to improving health outcomes for women, particularly in regions with high infertility rates like the Gulf. Here's an overview of what Diagnio brings to the table:

Features

1. Objective Data from Biosensors: The app utilizes advanced biosensor technology to collect accurate, real-time data on hormonal levels. This objective data is important to understanding individual health and fertility statuses.
2. Personalized Analytics through Algorithms: Diagnio applies sophisticated algorithms to generate personalized analytics using the data collected. These insights consider each user's makeup, providing tailored health and fertility information.

3. Actionable Advice via Online Consultations: Beyond data and analytics, Diagnio connects users with healthcare specialists for online consultations. This feature allows women to receive expert advice and actionable steps based on their diagnostic results.

Benefits

Enhanced Accessibility: Diagnio removes barriers to accessing healthcare by offering home-based or in-clinic diagnostics, making it easier for women to monitor their hormonal health and fertility.

Informed Health Decisions: The personalized analytics and advice provided by Diagnio empower women to make informed decisions about their health and fertility planning. This level of insight can lead to better results and a more proactive approach to healthcare.

Privacy: Diagnio's platform ensures user privacy while offering the convenience of managing health diagnostics from home. This combination of privacy and convenience is precious in areas where cultural sensitivities or logistics issues might restrict access to traditional healthcare.

Integration of Technology: Diagnio stands at the forefront of health tech innovation by using biosensors and AI-driven analytics. This technology integration improves the accuracy of health monitoring and sets a new standard for personalized healthcare solutions.

Global and Regional Impact: while Diagnio has a focused ambition in the Gulf market, its scalable solution has the potential to impact women's health worldwide. By addressing the high infertility rates in the Gulf, Diagnio not only contributes to regional health improvements but also demonstrates a model that can be adapted worldwide.

Diagnio's method of women's health diagnostics proves the power of merging tech and healthcare. By providing data, personal analytics, and professional advice, Diagnio helps women take control of their health and fertility in a way that was previously thought impossible. This platform plans to do its best to improve women's healthcare outcomes, especially in areas with urgent healthcare needs.

Milestone

Diagnio has hit a new milestone with the recent approval of its saliva-based testing kit by the Abu Dhabi Department of Health. This was quite a super moment for the company and the advancement of women's health diagnostics. This is not just a testament to Diagnio's wonderful approach to healthcare but also an important step forward in making women's health diagnostics more accessible, fast, and affordable.

The journey to this spot has been perseverance, collaboration, and commitment to meeting society's needs. Diagnio's focus on improving women's health through accessible diagnostics solutions has now been recognized and validated by a top regulatory body. This recognition has paved the way for the company to bring its groundbreaking technology to the market.

Diagnio's regulatory approval story began with a conversation between the company's representatives and Hesham Husni Abuasi. The conversation was on the importance of building strong relationships and navigating the legal, and regulatory scene with determination. Though filled with problems and learning opportunities, the journey shows the resilience and dedication of the Diagnio team to making a positive change in the healthcare industry.

With this regulatory go-ahead, Diagnio is set to make positive contributions to women's healthcare. The company's saliva-based testing kit is a major advancement in the field because it offers a beautiful solution for monitoring hormonal health and fertility for women. This development is beneficial for the Gulf region, where Diagnio has concentrated its efforts and where there is a pressing need for such healthcare innovations due to the high rate of infertility.

The Abu Dhabi Department of Health's approval indicates Diagnio's commitment to providing high-quality healthcare solutions and its potential to radicalize women's health diagnostics.

As Diagnio prepares to roll out its saliva-based testing kit to the community, there is anticipation for the positive changes it will bring to women's health management.

The achievement is a foundation for Diagnio's future endeavors in improving women's health. With a team obsessed with making women's health diagnostics quick and affordable, Diagnio is just beginning. But this is a wonderful step in the right direction for the company and women's health diagnostics field.

The healthcare community and Diagnio's supporters are waiting for the general adoption of its solutions and the future advancements the company will bring to the industry.

Case Study - CLINICY: Creating Digital Accessibility and
Efficiency of Healthcare Services

Clinicy - Offers Cloud-based services to Medical Institutions and Patients

A Saudi Arabia-based health tech start-up named Clinicy made noticeable strides in the healthcare technology sector in January 2024 by securing a seven-figure (USD) Series A funding round the same month. The round was led by Middle East Venture Partners (MEVP) and structured by Gate Capital, with additional participation from existing shareholders, including Kafou Group and Fadeed Investment. This funding is one of the largest investments in the Kingdom's health tech sector, and it is an apt reflection of the growing interest and confidence in digital healthcare solutions in the region.

Clinicy was founded in 2017 by Prince Mohammed Bin Abdulrahman Abdullah Al Faisal, Abdullah bin Sulaiman Alobaid, and Saud bin Sulaiman Alobaid. Its aim is to revolutionize the administrative aspect of healthcare through a cloud-based management system that automates important clinical processes like patient onboarding and retention. This directly solves the administrative inefficiencies that have long plagued the healthcare system.

The recent investment will help further the company's mission to expand its technological reach and improve its capabilities. The funds will also support the development of Clinicy's proprietary 'Interconnected HealthTech Ecosystem.' The initiative aims to fast-track the evolution of the health tech sector. Funny or not, this seems to align with Saudi Arabia's Vision 2030, which focuses on transforming healthcare efficiency and accessibility across the Kingdom.

Prince Mohammed Bin Abdulrahman Abdullah Al Faisal is the CEO and Co-Founder of Clinicy. While expressing his gratitude for the investment, he noted the importance of supporting the company's vision, which is to improve the quality of engagement between medical institutions and patients. He stressed that Clinicy is already making a tangible impact on healthcare in Saudi Arabia, enhancing digital experiences for over one million patients across the Kingdom. This latest investment will further advance Clinicy's expansion efforts and deepen its commitment to making healthcare more accessible and user-friendly.

Walid Mansour, the co-chief Executive at MEVP, pointed out the racing adoption of technology in Saudi Arabia's healthcare sector and attributed this to its digitally savvy population, government standards, and growing competition among medical service providers. He also stated that Clinicy's cloud-based platform helps clinics and medical centers offer online services, run cost-effectively, and continuously improve healthcare standards.

Munther Hilal, Gate Capital's chief executive, added that Clinicy has the potential to drive digital transformation in the Kingdom's health tech sector. This investment marks Gate Capital's first venture into the Saudi market, and it quickly followed the recent establishment of its Riyadh offices. Hilal showed enthusiasm for the Kingdom's myriad opportunities and a commitment to supporting and nurturing them in the future.

Clinicy's innovative solutions include its customized patient experience platforms, designed to solve problems like high patient 'no-show' rates and administrative inefficiencies that cost the Saudi healthcare industry over SAR 3 billion annually. With Clinicy, medical institutions have seen a 75% reduction in missed appointments, which shows the platform's effectiveness in improving patient healthcare delivery.

Business Model

Clinicy has a business model that uses cloud-based technology to reduce the administrative processes of healthcare systems. This model dwells on the automation of expedient clinical operations like patient onboarding and retention. It also plans to address the prevalent administrative inefficiencies within the healthcare sector. Here's a breakdown of Clinicy's business model:

Revenue Streams

- Subscription-Based Model: Clinicy operates a subscription-based model whereby medical institutions pay a recurring fee to access its cloud-based management system. This model provides Clinicy with a steady stream of income while offering customers continuous access to the platform's services.

Customized Solutions: Clinicy offers customized solutions specifically tailored to the needs of health service providers, which generates additional income via premium services.

Value Proposition

Operational Efficiency: Clinicy's platform solves the problems of high administrative costs and inefficiencies in patient management. By automating onboarding, booking, and follow-up tasks, Clinicy helps health practitioners reduce no-show rates and improve healthcare operations.

Improved Patient Experience: The platform also improves the patient's experience by offering more streamlined communication and engagement tools. This helps retain patients and attract new ones through positive reviews.

-Data Analytics and Insights: Clinicy provides valuable analytics and insights through its cloud-based system. This helps healthcare providers make informed decisions, leading to better patient care and optimized clinic operations.

Market Focus

- Saudi Healthcare Sector: Clinicy's initial focus on the Saudi market made it tap into a sector worth SAR 7.2 billion. This was done to lend a helping hand to the racing evolution of health tech in alignment with Vision 2030's goals. The focus on the market in a specific geographical location made Clinicy tailor its offers to meet local regulatory requirements and health needs.

Interconnected HealthTech System: The recent funding round was used to accelerate the development of an interconnected healthtech system. This ambitious project intends to broaden Clinicy's technological reach, offering an integrated platform that connects patients, health practitioners, and even regulatory bodies, thus improving accessibility to healthcare.

Strategic Partnerships

- Collaboration with Investors: Clinicy enjoys strong support from Middle East Venture Partners (MEVP), Kafou Group, and Fadeed Investment. These collaborations have positioned Clinicy for growth as they provide financial resources and guidance in navigating the health tech scene.

Clinicy's business model dwells on technological innovation and operational efficiency. With these, the start-up is ready to revolutionize the healthcare administrative space in Saudi Arabia and probably beyond. Here are some characteristics of Clinicy's platform:

Key Features

1. Software-as-a-Service Health Information System: The platform's SaaS system offers health service providers a comprehensive collection of tools for managing patient information, streamlining workflows, and improving medical efficiency.

2. Health Management Information System: Clinicy's HMIS facilitates collecting, analyzing, and reporting health information. This helps health institutions make data-driven decisions for better healthcare delivery.

3. Revenue Cycle Management: This feature helps health practitioners optimize their revenue processes, from patient registration and insurance verification to billing and collections.

4. Client Relationship Management (CRM): Clinicy's CRM system improves communication and engagement with patients, providing proof of satisfaction and exacerbating retention rates.

5. Invoicing Process Compliance: Clinicy ensures that Its invoicing process fully complies with Zakat, Tax, and Customs Authority standards. For medical professionals in the Kingdom, this provides peace of mind.

6. Automation of Administrative Tasks: The platform automates approximately 70% of employee tasks, such as service reminders and booking updates, reducing manual administrative work and allowing staff to focus on patient care.

Benefits

- Increased Efficiency: By automating clinical processes, Clinicy administrative inefficiencies and helps medical service providers offer faster and more reliable patient services
- Improved Patient Experience: The platform's CRM and automated communication tools improve the patient's experience by providing timely reminders, updates, and personalized engagement, leading to higher satisfaction and loyalty.
- Financial Growth and Stability: Clinicy's revenue cycle management feature helps ensure that health service providers maximize their revenue potential, achieve gross profitability, and are on track for net profit by 2025.
- Reduction in Missed Appointments: Health institutions using Clinicy have recorded an average 55% reduction in patient no-shows, with some institutions experiencing up to an 85% decrease. This not only improves their efficiency but also contributes to better patient outcomes.
- Data-Driven Decision Making: The health management information system offers valuable insights into patient care and operational performance. This enables medical institutions to make informed decisions that enhance healthcare delivery.
- Regulatory Compliance: With a thorough invoicing process that is compliant with local regulations, Clinicy ensures that health service providers meet all legal requirements.
- Scalability: Clinicy's platform is designed to meet the needs of a growing customer base, thus supporting the rapid development of Saudi Arabia's health tech space.

Partnerships

Clinicy has been expanding rapidly since its successful $5 million Series A funding round. Led by Prince Mohammed Al-Faisal and co-founders Abdullah bin Sulaiman Alobaid and Saud bin Sulaiman Alobaid, the start-up plans to achieve sevenfold growth by the last quarter of 2024.

Clinicy is currently valued at over SR7.2 billion. This valuation places it on a pedestal as one of the drivers of digital innovation in the area of improving accessibility to healthcare services.

With a vision aligned with Saudi Vision 2030 and collaborating with the Ministry of Health, Clinicy aims to bring about digital reform in the healthcare sector.

Prince Mohammed Al-Faisal, CEO and co-founder, stressed the company's role in creating digital accessibility and efficiency in healthcare services, thus showcasing the commitment at all levels - from governmental to institutional and single clinic - to induce positive patient results. This collaborative approach is vital for transforming the Kingdom's healthcare scene.

The Saudi government's allocation of SR214 billion ($57 billion) for health and social development in the 2024 budget shows the emphasis on investing in the healthcare sector in recognition of the pivotal role of innovative health-tech companies like Clinicy in supporting this growth.

Clinicy's platform offers a wide range of products for medical institutions. These include Software-as-a-Service Health Information Systems, Health Management Information Systems, Revenue Cycle Management, and Client Relationship Management. They all adhere to Zakat, Tax, and Customs Authority standards. In addition to tackling the challenges within Saudi Arabia's healthcare sector, Clinicy wants to introduce business-to-customer offerings in 2024 to improve patient experiences further.

Close collaboration with health practitioners ensures that updates to the Clinicy platform simplify operations and remain beneficial.

Clinicy has made a tremendous impact on reducing no-show rates. The platform has also automated approximately 70 percent of employee tasks, which has decreased medical centers' manual administration hours.

In 2024, Clinicy plans to diversify its income streams by offering direct business-to-consumer (B2C) services, thus moving past subscription-based partnerships with medical institutions. The company is already grossly profitable, but it still plans to achieve net profit by 2025 and focus on expansion within the Kingdom to tap into the wide domestic opportunity.

Clinicy's programs and financial capabilities have made it a leader in transformational healthcare delivery in Saudi Arabia. Its focus on digital innovation, collaborative efforts with the healthcare sector, and alignment with national development goals show its potential to have an even greater impact on Saudi Arabia's health tech space.

Other Health Tech Start-ups are:

1. Webteb
2. Healathand
3. Yodawy
4. Okadoc

CHAPTER 4

PropTech – Innovative Digital Solutions for the Real Estate Market

Proptech is a term that is short for property technology. PropTech has wholly transformed the real estate sector in the Arab region via the imbibing of innovative e-solutions into the property market to improve efficiency, transparency, and general customer experience. This sector uses technologies like artificial intelligence, big data, virtual reality (VR), and blockchain to narrow down and transform traditional real estate processes, from property search and transactions to management and investment.

Features of Proptech

1. Digital Platforms for Property Transactions: An online marketplace that connects buyers, sellers, and renters and offers a smooth and efficient property search experience.

2. Virtual Tours and Augmented Reality: VR and AR technologies allow potential buyers and tenants to view properties virtually online. One can look at how a property is designed and even walk through all the rooms virtually, thus preventing the need for physical visits.

3. Blockchain for Secure Transactions: Blockchain technology ensures secure and transparent property transactions. It reduces the risk of fraud and streamlines the buying, selling, and leasing process.

4. AI and Big Data: Artificial intelligence and Big data analytics give insight into market trends, pricing, and customer behavior. This enables more informed decision-making for investors, developers, and consumers.

5. Smart Home Technology: Integrating IoT (Internet of Things) devices and smart home technology into residential and commercial properties has increased comfort, security, and energy efficiency.

6. Property Management: Cloud-based solutions for property management help land owners and property managers automate tasks like tenant screening, rent collection, maintenance requests, and financial reporting.

Benefits of Proptech in the Arab World

1. Increased Efficiency: Automating traditional real estate processes reduces the time and effort spent on transactions, property searches, and property management.

2. Increased Transparency: Digital platforms and blockchain technology provide a transparent record of transactions, ownership, and property history, which builds trust among buyers, sellers, and investors.

3. Improved Accessibility: Online platforms and mobile apps make it easy for users to access property listings, market data, and investment opportunities anywhere, at any time.

4. Data-Driven Choices: AI and big data analytics help stakeholders make accurate predictions about market trends, pricing strategies, and investment returns.

5. Personalized Experiences: Advanced technologies like AI can prepare bespoke property searches and recommendations based on individual preferences, thus improving the customer journey for buyers and renters.

6. Reduction in Cost: Digitalizing real estate processes can reduce the need for physical paperwork, in-person viewings, and manual administration, leading to cost savings for developers, property managers, and consumers.

Property technology in the MENA region is currently driven by a young, tech-savvy population, government schemes targeted at diversifying the economy, digital transformation, and a growing realization of the need for more sustainable and efficient real estate practices. Property technology is prepared to play its role in shaping the future of the real estate industry in the region and offering new opportunities for innovation and growth.

Case Study - HUSPY: Simplifying Home Finance

Huspy - Transforming the Home-Buying Process

Huspy is a top company in the property technology space in the Arab world, known for its extraordinary approach to simplifying the home buying and mortgage process. Jad Antoun and Khalid Ashmawy established Huspy with a vision to narrow down the complicated procedures of real estate transactions. The company uses technology to offer more transparent, efficient, and user-friendly services. Let's look at Huspy's key features, benefits, and impact on the proptech landscape.

Features of Huspy

1. Virtual Mortgage Application: Huspy has a virtual platform where users can apply for mortgages. This reduces the paperwork and the need for in-person visits to banks or financial institutions.

2. Real Estate Listings: Huspy offers a curated selection of real estate listings. Users can search through the available properties to find one that meets their location, price, and size criteria.

3. Automated Mortgage: One outstanding characteristic is Huspy's ability to offer users a comparison of different mortgage products from various lenders, helping them find the most competitive rates and terms.

4. End-to-end Support: Huspy offers customer support throughout the home purchase procedure - from property search and mortgage application to closing the deal. This includes providing users with expert advice and negotiation support.

5. Transparent Processes: The platform has a super transparent process that provides users with clear information about fees, rates, and everything else. It demystifies the complexities of buying a home.

Benefits for Users

1. Convenience: Huspy, by digitizing the whole home buying process and documentation, allows users to manage most of the process from the comfort of their homes.
2. Time Savings: Huspy's integrated processes and automated tools save users time, speeding up property searches, mortgage applications, and home comparisons.
3. Cost Savings: With access to a wealth of mortgage options, users can identify the most cost-effective solutions, thus saving themselves thousands over the life of a mortgage.
4. Expert Guidance: Huspy gives users access to real estate and mortgage experts, who help them make informed decisions at every step of the process.
5. Increased Accessibility: The platform makes the real estate market accessible to a wider audience, including first-time buyers and those who may have found traditional processes intimidating or complicated.

Impact in the Arab World

Huspy's emergence as a proptech company in the Arab world shifted towards digitalizing the real estate/property sector. Its innovative and creative approach addressed many traditional pain points associated with home buying, like lack of transparency, inefficiency, extensive documentation, and sky-high costs. Thus, Huspy refines the home-buying experience for individuals and contributes to the modernization and growth of the real estate market in the region.

The success of Huspy and similar proptech start-ups in the Arab world further encourages innovation and investment in the sector, promising to bring about more advancements in real estate transaction conduct.

This aligns with broader digital transformation goals across many Arab countries and is part of global efforts to diversify economies and improve service sectors.

Business Model

Huspy's business model integrates technology with the real estate market to change how homes are bought and financed in the Middle East and Europe. This model provides a comprehensive platform for property discovery and mortgage financing. It uses its relationships with banks to offer competitive financing options to its users.

Here's a closer look at Huspy's business model:

Key Components

1. Mortgage Aggregation and Financing: Huspy acts as a mortgage aggregator, partnering with UAE banks to offer vast financing options. By combining offers, Huspy allows its customers, whether residents or non-residents, to compare and select the most competitive interest rates, starting from as low as 4.24% for a fixed three-year basis.

2. Technology-Driven Platform: Huspy's business model is based on its innovative technology platform, which simplifies the home buying and financing process. The platform helps customers quickly find suitable properties and secure finances, streamlining what has traditionally been a complex and time-consuming process.

3. Free Mortgage Services: Huspy offers free mortgage services to home buyers, which is a significant value proposition. This approach most likely attracts a larger user base since it removes

upfront costs for customers seeking mortgage advice and financing solutions.

Revenue Streams

1. Referral Fees from Banks: Huspy earns referral fees when it links its customers to banks for mortgage financing. Banks benefit from the inflow of qualified leads, while Huspy earns a commission for every successful loan disbursement.
2. Partnership Fees: Due to its collaboration with different real estate entities, Huspy charges partnership fees for listings and promotions on its platform. These include featured listings for developers or promotional campaigns for banks.
3. Ancillary Services: Huspy also monetizes other services related to home buying, like property valuation, legal services, or insurance, either directly or through partnerships with the service providers.

Market Position and Expansion

Huspy has established itself as the UAE's largest mortgage platform and is still expanding. Its success, highlighted by processing over AED 1 billion in home financing in a single month, shows the widespread demand for innovative proptech solutions. Huspy's growth is further facilitated by the UAE's robust economic environment, supportive start-up space, and high demand for property ownership.

Future Growth

Huspy's future growth may involve expanding its tech offers, penetrating new markets beyond the UAE and Europe, and diversifying its services to include more aspects of the real estate transaction process.

Huspy is prepared to maintain its leadership in the proptech sector and redefine the future of real estate and mortgage financing.

Case Study - BAYUT: AI-assisted Property Search

Bayut - AI-Powered Real Estate Portal

Bayut is another pioneering force in the MEA proptech industry, especially in the United Arab Emirates. Imran Ali Khan and Zeeshan íuAli Khan had the vision to simplify the property search process, leading to Bayut's establishment. It is currently one of the top property portals in the region.

It boasts a comprehensive database of real estate listings that cater to various needs—from rentals to sales of both commercial and residential properties.

The platform has BayutGPT, an AI-powered property search assistant with a user-friendly interface. This has completely transformed how individuals and families find their next home or investment opportunity.

By lacing the real estate market procedures with cutting-edge technology, Bayut has improved the property search experience and made it more accessible, efficient, and user-specific across the UAE and beyond.

Business Model

Bayut has a model that falls within the dynamic property technology space in the United Arab Emirates and the broader Middle East region. It is a multifaceted business model that caters to the needs of both property seekers and providers. Its business model revolves around connecting buyers, sellers, and renters of real estate with the best possible matches facilitated by a sophisticated online platform. Here is an exploration of the critical components of Bayut's business model:

Revenue Streams

1. Subscription Fees: Bayut offers subscription packages for real estate agencies and developers. The packages include listings and enhance visibility on the platform. These subscriptions come in different prices and features, so they are flexible based on the size and needs of the agency or developer.
2. Featured Listings and Advertisements: Bayut also allows sellers and agents to feature their listings on the site. Such placement attracts higher visibility and, perhaps, quicker sales or rentals. Bayut also generates revenue through targeted adverts displayed on its platform to catch the substantial traffic of property seekers.
3. Lead Generation Fees: Bayut charges a small fee for generating leads by acting as a bridge between property seekers and sellers. This is valuable for real estate agents and developers looking for specific potential buyers or tenants.
4. Value-Added Services: Bayut offers various other value-added services, such as market reports, property price trends, and investment insights, some of which can be monetized.

Competitive Advantage

1. User Experience: Bayut stands out because of its user-friendly interface and innovative tools like BayutGPT. These offer a smooth user experience, thus retaining and attracting new users.
2. Comprehensive Database: Bayut has an extensive listing of properties for sale and rental. This ensures that property seekers have access to a wide range of options, from affordable apartments to luxury villas in different regions of the UAE.

3. Market Insights: The platform provides insights into the real estate market, benefiting both property seekers and sellers by guiding their decisions with data-driven analysis.

Market Positioning

Bayut has positioned itself as a leader in the UAE's proptech sector by continuously innovating and adapting to market needs. The company's emphasis on technology-driven solutions for home search and home buying/lease transactions appeals to the region's tech-savvy audience. Their approach also caters to the rising demand for convenience, efficiency, and reliability in property transactions. As a comprehensive property portal, Bayut has several features and benefits that cater to broad groups of users.

Features

1. Extensive Property Listings: Bayut has various property listings, ranging from apartments and villas to commercial spaces. This extensive selection ensures that users can find options that suit their needs and preferences, whether they want to buy, sell, or rent.
2. User-Friendly Interface: The platform's user-friendly interface simplifies the entire property search procedure. Users can use BayutGPT to engage conversationally to find properties, making the search experience more intuitive and less time-consuming.
3. Data-Driven Insights: Bayut provides users rich recommendations and insights for making informed decisions. Whether understanding market trends, identifying investment opportunities, or analyzing rental yields, Bayut equips its users with important information.

4. AI-Powered Tools: The platform's technology offers personalized property recommendations and answers to users' real-time questions.

5. Market Coverage: Bayut covers a wide range of locations within the UAE, including Dubai, Abu Dhabi, Sharjah, and so on. This broad coverage allows users to access market information regardless of their geographical preferences.

6. Trust and Reliability: Bayut is known for its trustworthiness and reliability. Users can confidently navigate the platform, knowing that listings are verified and up-to-date. This is super important in a market where transparency is highly valued.

7. Community and Support: Bayut has a network of users and it offers them support via its customer service schemes and online resources. This community aspect provides a layer of guidance and assurance for users in the real estate market space.

Expansion and Growth

Bayut's major focus is the UAE market. However, its business model and technological infrastructure allow scalability and expansion into new markets within the Middle East and beyond. It is certain that Bayut can replicate its success in other regions if it dares to tap into the global interest in such proptech solutions. It can do this by pushing out its successful platform and adapting to local market niceties. The platform's business model, which is centered around technology, customer service, and its broad market coverage, made it a top player in the digital transformation of the UAE's real estate sector.

Its continued investment in innovation and user experience keeps Bayut at the forefront of the proptech industry, driving the future of real estate in the region and beyond.

CHAPTER 5

Mobility Tech - Shaping the Future Mobility Landscape

There is a tech space known as shared mobility technology. It includes different services like sharing rides, renting cars, shopping for groceries, and ordering cab models. This tech space has moved upward in the MEA region, especially after the global success of Uber and Lyft.

This sector of tech has produced several unicorns, but it still has the potential to expand even further. According to Grand View Research Inc., the shared mobility market in the MEA region can grow at a compound annual growth rate (CAGR) of 18.4% from 2022 to 2030. The annual market should also see a 16.9% increment in these eight years.

This increase is possibly due to factors like increasing urbanization in the MEA region, rising demand for cost-effective and convenient transportation options, and growing awareness and acceptance of shared mobility services among the population. Improvements in technology and the digitalization of services have also made it easier for consumers to use shared mobility platforms, making the sector expand even more.

The governments in the MEA have not been asleep in this rising tech scene. They have been involved in implementing regulations that help develop shared mobility services. These legal and regulatory frameworks have helped promote safety, reliability, and fairness for users and service providers, creating an atmosphere that is super conducive to the growth of the shared mobility market.

The shared mobility market is still evolving in the MEA region. It offers several opportunities for innovation and investment. Companies operating in this space are always exploring new models and technologies, such as electric and autonomous vehicles, to improve their offerings and meet consumers' insatiable needs. The market will continue to expand in the coming years. As a result, the shared mobility sector in the MEA region is a dynamic and promising area for further development and growth.

The Middle East is quickly emerging as a viable player in shaping the future of mobility, driven by its increasing population, urbanization, and technological investment. The region's commitment to transforming everyday transport into a super convenient, safe, and eco-friendly experience is seen in adopting new technologies like electric and autonomous vehicles and even cutting-edge urban air mobility solutions like electric vertical takeoff and landing (EVTOL) aircraft.

The Gulf States, in particular, are at the helm of this transformation as they aim to improve living standards and push the region towards a greener future through mega projects that redefine average transportation. The Middle East's preference for electric and autonomous vehicles stands out in this transportation revolution. The governments of UAE and Saudi Arabia have made plans to increase these advanced transport styles to eradicate reliance on fossil fuels and ease the quality of urban air. For instance, Dubai plans for 25% of all transportation to be autonomous by 2030. The region is taking a proactive stance in embracing the future of mobility.

Saudi Arabia's move into electric vehicle (EV) manufacturing started with the launch of Ceer, the first Saudi Electric Vehicle brand in collaboration with international tech and automotive giants. This action was a reflection of the region's industrial ambitions. This move aligned with significant agreements to establish EV manufacturing facilities and showed a commitment to fostering an EV ecosystem despite the traditionally low fuel costs in the region.

Developing intelligent transportation systems across cities like Doha, Abu Dhabi, and Riyadh is another thing that makes the region stand out. These smart systems are to improve traffic flow, reduce traffic jams, and improve road safety through data-driven insights and analytics. Initiatives such as this show a broad strategy to reduce the adverse effects of traffic jams and promote a shift towards public transportation and shared mobility services, further supported by regulations like the UAE's Vertiports framework.

It is interesting to know that Irish companies are actively contributing to the Middle East's mobility evolution by offering innovative solutions across different aspects of transportation. From CitySwift's AI-powered optimization of bus networks to Cubic Telecom's connectivity solutions for the next generation of vehicles, Ireland is leading the pack toward addressing global mobility problems. Also, initiatives like the Future Mobility Campus Ireland (FMCI) show Ireland's role as a hub for mobility innovation and, as such, provide a sort of collaborative testbed that will further air and land mobility technologies.

As the Middle East moves closer to a futuristic mobility space, there is greater synergy between regional ambitions and global innovation, especially from Ireland, as all nations have chosen to redefine transportation norms. This approach will not only accelerate the transition to sustainable mobility solutions but also align with the greater goals of reducing carbon emissions and diversifying oil-dependent economies, thus charting a course for a more sustainable and technologically advanced future in transportation.

Case Study - CAREEM: SuperApp for the Greater Middle East

Careem - the Everything App

The first time Careem was launched, it was a simple ride-hailing service. It then evolved to become an all-encompassing tech platform that is simplifying the lives of users through an app called the 'Everything App'. The company was co-founded by Mudassir Sheikha and Magnus Olsson in 2012. Currently, it is a beacon of entrepreneurial success in the Middle East and North Africa. Under the leadership of its co-founder, Mudassir Sheikha, Careem has achieved so much success since its inception and expanded its services beyond mobility to include over a dozen other digital services.

The company's 'Chapter 2' phase started in 2019 1ith the acquisition of its ridesharing business by Uber. This paved the way for more expansion and innovation. Later on, the UAE-based tech and investment conglomerate e& acquired a majority stake in Careem Technologies. This was a sign that a new era of growth and diversification was underway.

Co-founder Sheikha talked about Careem's commitment to its original values and purpose as they get the credit for the company's enduring success. He insinuated that by staying true to its mission, Careem has facilitated the creation of almost three million earning opportunities and has equally pushed the boundaries of possibilities in the region, thus inspiring others to dream big. The partnership with e& and Uber further strengthened Careem's capabilities and enabled it to reach more people and expand its services beyond mobility alone.

Sheikha is quite optimistic about the potential of Careem's 'Everything app' in the Middle East as it addresses most of the region's daily problems via digital solutions. As no dominant players are offering the convenience and value that Careem provides, there is a gigantic opportunity for growth and impact. While reminiscing on the past decade, Sheikha acknowledged the importance of learning from mistakes and the value of hiring trustworthy leaders to delegate responsibilities effectively. This approach has made the company scale and adapt as trust and effective leadership contribute to long-term success.

Innovative Business Model

Careem went beyond traditional ride-hailing by adding additional services like food delivery (CareemNOW) and digital payment (CareemPAY). The expansion was one half of the company's vision to become the region's leading Super App with many services under one umbrella. This move was not just a diversification plan but a plan to create a more sustainable and scalable business model.

Acquisition by Uber

In March 2019, Uber bought Careem for $3.1 billion. This took place barely six years after the company was set up and the acquisition was one of the biggest tech transactions in the Middle East. This acquisition was quite beneficial to Uber as it allowed them to consolidate their operations in a region where Careem had a stronger base. For Careem, the acquisition gave it access to Uber's global resources, technology, and expertise. This was useful for further expansion and allowed the company to proffer more services.

Business Model

Careem is, indeed, a trailblazer in the ride-hailing and tech space. The company has been setting the pace in this area since it was set up. From the time it was launched as a ride-hailing service till now, Careem has expanded to include several digital services in line with its aim to become the 'Everything App' for the region. This is a breakdown of Careem's model:

1. Ride-Hailing Services: Careem began as a simple ride-hailing service where users booked rides through an app. It had different vehicular options for different customer preferences and budgets. This remains the major part of their business, and they generate income through a commission model, in which Careem takes a percentage of the transport fare from each ride.

2. Super App Strategy: The company expanded beyond ride-hailing and ventured into multiple digital services, including food delivery, grocery shopping, and digital payments, all through its Super App. This made the company a superb platform where a suite of services was offered to meet the everyday needs of its users and provided additional income.

3. CareemPAY: CareemPAY is the e-payment platform in the Super App. It carries out in-app payments for Careem's services and peer-to-peer transactions. This extends the company's income model to financial technology as it then offers convenient payment solutions to its wide audience.

4. Collaborations: Careem partners with various businesses and service providers to enrich its Super App. These relationships help the company to offer a lot of services and generate additional income through referral fees, partnerships, and promotional collaborations.

5. Data: Careem has a vast amount of data as it uses analytics to make decisions, engage in personalized marketing, and improve its users' experience. This is not a direct revenue source, but data analytics enhance operational efficiency and service offers, thus boosting revenue.

6. Expansion into New Markets: Expansion into new geographical markets and service areas allows the company to tap into new customer bases. It does this by carefully selecting markets with growth potential and tailoring their services to meet local needs.

7. Corporate and Business Solutions: Careem offers customer-centered solutions for businesses. These include corporate travel, logistics, and delivery services. This B2B business model provides the specific needs of the companies, which is a reliable service for generating additional income.

Challenges

Careem's model has problems like legal hurdles, competition, and the need for constant innovation. The company's love for adaptation can be seen in its expansion into new services and markets. This also shows its resilience and commitment to meeting the changing needs of its users. Careem's business model reflects a deep understanding of the MEA's market dynamics and consumer behavior. Careem remains a top player in the Middle East's tech ecosystem with its continuous innovation and expansion of its service offerings as well as a clear vision for future growth and development.

Features

As stated previously, since its inception, Careem has gone past its original ride-hailing service to add on a wide number of digital services as it aims to become the 'Everything app' for the Middle East and North Africa region. This transformation shows Careem's response to the changing needs of customers and the dynamic digital space in the MEA. Some key features of Careem's business model are:

1. Different Mobility Solutions: Careem started as a ride-hailing service, offering various vehicular options to fit different customer needs like the economy, business, and family rides. Then, it expanded to include options for car and bike rentals for users looking for short-term vehicle access without the commitment of ownership.

2. Super App Strategy: Careem has offered multiple services under one platform since transitioning into a super app. It has a culmination of multiple services like food delivery, groceries, payment services, and its traditional ride-hailing service. This strategy longs to simplify users' lives by offering a one-stop solution for their daily needs. With CareemPAY, the digital payment platform within the app facilitates smooth payment transactions for different services.

3. Focus on Local Needs and Regulations: Careem tailors its services to meet the specific needs and preferences of users in different MENA region countries while considering cultural norms and local regulations. The company also works closely with local governments to ensure compliance with transport and digital service regulations, thus fostering a cooperative relationship that benefits both parties.

4. Technology and Data Analytics: Careem uses advanced technologies like artificial intelligence and machine learning to optimize its services. It also uses data analytics to continuously improve its services and customer experience based on user feedback and behavior analysis.

5. Community and Social Impact: By providing flexible earning opportunities for captains (drivers), Careem has contributed significantly to job creation in the region. The company also values sustainability, which is why it incorporated electric

vehicles into its fleet and promoted environmentally friendly transportation options.

6. Expansion: Through its several partnerships and business relationships, such as Uber's acquisition and collaboration with e&, Careem has increased its service portfolio and technological abilities. The company also keeps on exploring new service areas and technologies, including the potential introduction of autonomous vehicles and urban air mobility solutions in the future.

Expansion Strategies

1. Location Expansion: Since its first launch in Dubai, Careem has expanded its services across the Middle East, North Africa, and South Asia. By identifying and gaining access to new markets with a demand for ride-sharing services, Careem has used the first-mover advantage in many regions and established a strong presence ahead of global competitors.

2. Service Diversity: Careem's evolution into a super app shows its move to diversify services. Apart from ride-hailing, Careem now offers food delivery, payments, and logistics services. This allows it to meet greater customer needs, making the app a daily one for its users.

3. Technology Investment: Technology investment has been central to Careem's expansion. It continuously upgrades its platform to improve user experience, retaining customers and attracting new ones. Its investment in AI and machine learning technologies, exemplified in the development of CareemGPT, further places the company at the helm of innovation.

Partnership Strategies

1. Collaboration with Governments: Careem collaborates with several government entities, aligning its services with regional development plans, such as Dubai's goal of 25% autonomous transportation by 2030. These collaborations ensure regulatory compliance and support the company's integration into the global urban mobility system.

2. Acquisitions: Uber's acquisition of Careem was a partnership relationship that gave Careem access to Uber's global resources and expertise. Careem has also acquired or invested in several local start-ups to expand its service offerings and technological abilities.

3. Joint Ventures: Careem has entered into several joint venture agreements, like its recent partnership with Saudi Arabia's Public Investment Fund (PIF) and other entities, to launch new services specifically made for the regional market. These ventures allow Careem to use such local partners' strengths and market knowledge.

4. Corporate and Financial Partnerships: collaborating with financial institutions and corporate entities has helped the company introduce financial services within its app. An example is the CareemPAY digital wallet.

Looking Ahead

Careem has shown that it is committed to simplifying and improving the lives of people in the MEA region. This is why it keeps expanding its services and looking for new business relationships. The company's mode of expansion and investment in technology and innovation, has placed it on top in the MENA region's digital scene. With the backing of strong partners like Uber and e&, Careem is ready for future growth opportunities and challenges.

Case Study - SWVL: Book your Ride App

SWVL - Book your Ride

Swvl is a mobility start-up that is based in Dubai. It was launched in 2017, and ever since then, it has kept growing to the point where it is now a top player in the global shared transportation sector. Mostafa Kandil, Mahmoud Nouh, and Ahmed Sabbah set up the company. The trio planned to transform the public transport space by offering tech-driven alternatives to traditional public transportation and private car ownership. Since it uses a tech platform, Swvl helps its customers book fixed-route shared bus rides through its mobile app, a more efficient, reliable, and cost-effective transportation solution.

Business Model

Swvl has a unique business model that is a middle spot between traditional ride-hailing services and public transportation. It uses a bus aggregation model to organize mass transit on fixed bus routes, allowing users to reserve bus seats through its app. This model fills in the market gap for affordable, reliable, and quality public transport options. The company operates a B2B and B2G model where it partners with corporate clients and governments to provide bespoke mobility alongside its B2C offers.

Features

1. Fixed Routes and Schedules: Unlike normal ride-hailing services, Swvl offers transport services on predetermined routes with pre-set schedules. It combines the convenience of ride-hailing with the efficiency of mass transit.
2. App-Based Booking: the Swvl app usually displays route options, schedules, and available seats so customers can easily book and pay for rides through the said app.

3. Dynamic Routing: Swvl uses certain routes based on demand and traffic conditions, improving efficiency and user experience.
4. Safety and Comfort: Swvl places a premium emphasis on safety and comfort with offers like live tracking, driver ratings, and well-maintained vehicles.

Expansion

At first, the company was only focused on the Egyptian market, but it has since moved its operations across different countries in the Middle East, Africa, and Asia. The company's method of scaling involves entering markets with substantial demand for efficient and affordable public transport options. Its recent focus is on narrowing its operations to markets in Egypt, Saudi Arabia, and the UAE as it gains profitability and sustainable growth. Swvl's partnership strategy includes collaborations with local transport providers, government entities, and corporate clients. These partnerships have helped the company to teach its services into the transport systems available before, expand its user base, and diversify its income.

Challenges

Swvl's journey has been laced with problems that include legal hurdles, operational dilemmas, and the impact of the Coronavirus pandemic on public transport. The company's adaptive tactics - shifting towards a B2B model and narrowing down its operations, have placed it as one brand with an optimistic future. Swvl is ready to continue its mission of transforming urban mobility. As cities all over the world seek sustainable transport solutions, Swvl's model could prove sacrosanct in shaping the future of mass transit, making urban mobility more accessible and environment-friendly.

Profits Milestone

Within the first 6 months of 2023, Swvl got an exceptional milestone by posting a net profit in its financial report. This miraculous turnaround was after a series of decisions aimed at narrowing down its operations, including divesting its subsidiaries in over a dozen countries and focusing only on Egypt, Saudi Arabia, and the UAE. The company also shifted towards a B2B model and reduced the number of employees as part of a restructuring.

Utterly contrary to the $161.6 million loss reported within the first 6 months of 2022, Swvl got a net profit of $2.1 million in the same time frame in 2023 despite a 49% year-on-year revenue decrease. This recovery is more evident in light of the operating profit of $13.4 million, compared to a $56.0 million operating loss in H1 2022. The major part of Swvl's income, which accounted for 73.7%, comes from sales of technology that helps clients plan their routes, operate fleets, or manage riders, while the remainder comes from its bus operations.

Egypt is still Swvl's largest market, contributing about 93% of the company's revenue. Saudi Arabia contributed 7%, while the UAE did not contribute to the company's income at all in this period. The cost of sales also fell by 61% to $9.3 million compared to the first six months of 2022.

Swvl's CEO, Mostafa Kandil, is confident in the company's new direction. He also stressed the creation of value for shareholders and the positioning for profitable growth in the high-revenue markets. The company reported a positive cash inflow of $2.2 million in the first six months of 2023. This was a major improvement from the operating outflows of $76.8 million in the first six months of 2022. Investors have welcomed this financial health, especially as Swvl's share price more than doubled after the release of the financial report.

The sale of one of its subsidiaries - Urbvan Mobility Ltd. to Kolors Inc., a top transport service provider in Latin America, for $12 million in cash, shows the company's focus on prioritizing higher revenue markets and reducing its operations. This decision is part of Swvl's broader strategy to adapt and thrive in a challenging economic environment plagued by currency devaluation in its primary market, Egypt, and a general reduction in investor appetite for startups.

With its return to profitability, Swvl has positioned itself for a new growth stage. Its ability to generate positive cash flow and reduce liabilities has restored investor confidence, which is clear in the over 100% rise in its share price after the announcement of its financial results. Swvl's move to a technology-focused B2B model shows a transformation from its original business model. The company's success in being profitable despite unfavorable market conditions shows the need for adaptability, focus, and operational efficiency in the competitive mobility sector.

The company's path to profit has been littered with challenges, including employee layoffs and the dissolution of subsidiaries across various countries. However, with a renewed focus on expansion, especially in the Saudi market, and a strategic rehiring plan, Swvl shows resilience and adaptability in the face of adversity. The company's successful move towards profitability, together with a meticulous expansion plan for 2024, symbolizes a promising future for Swvl as it keeps innovating and soaring.

CHAPTER 6

EdTech – Innovation in Education

The Educational Technology space in the MEA region has grown immensely, especially as a result of the COVID-19 pandemic's push towards e-learning. The region, which is known worldwide for its young population and high smartphone penetration, is indeed fertile for educational technology solutions.

Trends in Middle Eastern EdTech

1. E-Learning Platforms: the shift to online learning during the COVID-19 pandemic led to the adoption and use of virtual platforms. Schools, universities, and private institutions quickly moved online, using platforms like Google Classroom, Zoom, and local start-ups specifically tailored to the region's needs.

2. Language Learning and Supplementary Education Apps: Due to increased demand for bilingual or multilingual education in the region, language learning apps and platforms are now offering

supplementary education in English, Arabic, and other languages. This idea was introduced by parents who wanted to boost their children's skills outside the traditional classroom setting.

3. Increased Investment: The Middle East has also had a high investment rate in EdTech start-ups. Governments and private investors have now seen the potential of these companies to contribute to educational reform and economic diversification. Popular start-ups like the Noon Academy in Saudi Arabia and Almentor in Egypt have raised quite a lot of funds, and this shows investor confidence in the sector.

4. Focus on STEM Education: The Middle East is also currently placing an increased emphasis on STEM (Science, Technology, Engineering, and Mathematics) education. EdTech platforms now offer coding courses for kids, interactive science lessons, and mathematics tutoring as the region aims to nurture a future workforce skilled in these areas.

EdTech Challenges in the Middle East

1. Digital Divide: Urban areas in the Middle East have high internet connectivity, but rural and underserved communities lack access to reliable internet and digital devices almost constantly. As such, equitable implementation of EdTech solutions cannot be achieved.

2. Content Localization and Cultural Relevance: If EdTech platforms are to be successful, it is important to develop content that is culturally relevant and available in local languages. Several start-ups are facing the issue of tailoring educational materials to meet the differing needs of the Middle Eastern audience.

3. Regulatory Hurdles: Successfully scaling the legal hurdles in each country can be quite a hassle for EdTech companies, with variations in regulations across countries in the region. Establishing standards for e-education and ensuring compliance with local laws is expedient for these companies to thrive.

Government Initiatives Supporting EdTech

Governments across the Middle East are setting up several schemes to promote the EdTech sector and its potential to improve the quality of education.

For instance, Saudi Arabia's Vision 2030 includes plans to modernize the kingdom's educational space with substantial investments in digital infrastructure and partnerships with EdTech companies.

The UAE's Ministry of Education has equally launched several programs to inculcate technology into classrooms, including smart learning schemes and innovation labs. Egypt's Knowledge Bank, which is unarguably one of the world's largest digital libraries, is also among them, and it provides free access to students, teachers, and researchers.

EdTech Future in the Middle East

The Middle East's EdTech future looks promising. A cursory look will show its potential for growth with the continued inflow of investors, government support, and assistance from educational and research institutions. Also, people recognize the value of inculcating technology into learning. The sector's growth curve will most likely dwell on proffering solutions to present challenges like the digital divide and content localization. At the same time, novel solutions should be considered to personalize and improve the learning experience.

Certainly, the Middle East will soon be a top player in the global EdTech scene.

Case Study - NOON: Social eLearning Platform

Noon Academy – A Worldwide Social eLearning Platform

In 2013, Mohammed Al-Dhalaan and Aziz Al-Saeed set up the Noon Academy in Saudi Arabia. What began as a simple test-preparation platform has become one of the most prominent EdTech startups in the Middle East today. The Academy's mission is to empower students through social learning, and this is why it uses technology to create a platform that improves academic achievement and creates a supportive network of learners and educators.

Business Model and Innovation

Noon Academy's model is centered around a 'freemium' service – this means that it offers free access to live classes and interactive sessions with premium features available for a subscription fee. This model is open to a wide demographic, and this makes for accessibility while also generating income. What sets Noon Academy apart is its social learning method. Students can study together, compete, and ask questions in real-time during live sessions, making learning more engaging and collaborative.

Growth and Expansion

The Noon Academy was initially meant for and adapted to the Saudi market. But now, it has expanded its services to eight countries, including Egypt, Pakistan, and India, with a user base of over 10 million students. The platform has a plethora of subjects for the different academic phases. The Academy has partnered with educational/research institutions and content providers. This has enabled the start-up to localize its content and offer curricula relevant to each market.

Funding and Investment

Substantial investment rounds have supported Noon Academy's growth, showing investors' confidence in its vision and model. In 2020, Noon Academy received $13 million in a pre-series B funding round led by STV, and again in 2021, it received $26 million in a series B round. These monies have been useful in boosting its tech infrastructure, expanding its services, and penetrating new markets.

Challenges

Despite its success, Noon Academy's journey has been littered with challenges, including adapting to different educational systems and regulations across countries. Despite hurdles, it has set a standard of high-quality content and maintenance of student engagement in virtual learning environments. In addition, although the COVID-19 pandemic increased demand for online education, the global situation made Noon Academy quickly upscale and boosted infrastructure to accommodate the rise in users.

Impact

Noon Academy's impact on the MEA's education sector has been optimal. The company has provided an accessible, interactive learning platform for millions of students. It stresses social learning and community building, which have introduced a novel shift in education, where learning extends beyond the classroom. The Academy wants to increase its user base to 50 million students by 2023, and it plans to do this by focusing on international expansion and improving its platform with AI and machine learning technologies for personalized learning. With present investments in technology and content, Noon Academy is ready to continue its mission of transforming education in the Middle East and beyond.

Noon Academy's trajectory from a test-preparation platform to a top EdTech start-up in the Middle East shows the transformative power of technology in education. By focusing on social learning, accessibility, and scalability, Noon Academy has improved educational outcomes for millions of students and set a benchmark for standards in the global EdTech industry. The company promises to make a lasting impact on the educational space.

Factors for Success

In dynamic sectors like technology, education, and e-commerce, any enterprise's success is based on many factors that show the intricate details behind the success and act as guiding principles for new start-ups and enterprises looking to carve a niche in competitive markets. Some of these factors are;

1. Understanding Market Needs: Successful companies understand their target market's needs, pain points, and preferences. This knowledge helps them effectively tailor their products or services to address those needs. For example, Noon Academy's focus on social learning aligned with students' desire for a more interactive and engaging learning experience.

2. Innovation: This is the actual pillar of success in business today. Companies that consistently inculcate innovation into their offers and models usually manage to stay ahead of technological advancements and market trends, and the ability to do this is very vital.

3. Strong Leadership: Every company that plans to be successful must have visionary, decisive, and capable leaders who can inspire and mobilize the team toward a common goal. Such strong leadership will give the entire company the boost it needs to get through growth stages, problems, and major changes while remaining in line with its core values.

4. Quality and Value Proposition: If a business is to succeed, it is vital to give customers the best, most valuable products or services. Companies that succeed do so because they tend to provide solutions that are not just good but are perceived by their customers as better or more valuable than those of other brands.

5. Marketing and Branding: A marketing strategy that is well-executed and effectively communicates the business's value proposition to the target audience will positively affect a company's success. This is why it is essential to build a strong brand that resonates with customers and engenders loyalty.

6. Customer Focus: Companies that emphasize customer satisfaction usually perform better than others. Focusing on your customers includes providing excellent customer service, listening to feedback, and continuously improving customer experience.

7. Sustainability: The ability to efficiently scale operations while maintaining or improving profitability is a major plus for long-term success. Sustainability, whether in terms of environmental impact or business practices, is increasingly vital to customers.

8. Collaborations: Forging partnership relationships provides companies with access to new markets, technology, expertise, and resources. Such relationships can also improve innovation and offer a competitive advantage.

9. Talent Acquisition and Retention: Talent acquisition and retention are important because they promote innovation, maintain operational excellence, and further a positive organizational culture. Successful companies invest in their employees' training, development, and well-being.

10. Financial Management: Prudent financial management includes effective budgeting, cash flow management, and securing finances for growth schemes. These are all important steps to sustain operations and induce expansion.

Case Study - ARABEE: Learn Arabic Online

Arabee – App for Learning Arabic

In this digital age, language learning applications are popular for acquiring new languages as they offer flexibility, interaction, and accessibility. However, so far, the Arabic language, with its rich cultural heritage and complexity, has proved a seeming issue in virtual education. Arabee is a language learning program designed for non-native speakers to learn Arabic. It was developed to make learning Arabic fun and engaging, especially for children. It is a needed advancement in the EdTech space within the Middle East.

Background

An important observation drove the establishment of Arabee: Despite the significant time invested in studying Arabic, many children struggled to communicate effectively in the language. This predicament was further increased by the lack of engaging and interactive learning materials, which failed to capture the interest of young learners. After identifying these problems, Lenka Basweidan, the founder of Arabee, together with a team of teachers, parents, and professionals, set out to create a solution that would redefine learning the Arabic language.

Objective

The primary objective of Arabee was to develop a comprehensive, interactive, and enjoyable Arabic language learning platform. It aimed to:

1. Increase engagement and interest in Arabic language learning among non-native speakers, particularly children.
2. Provide a curriculum that aligns with international language standards and the UAE Ministry of Education.

3. Address the educational gap by offering an e-platform that supports progressive learning.

Strategy

Arabee's strategy has a couple of elements, which are:

1. Interactive and Hybrid Learning: This entails incorporating fun videos, sing-along songs, interactive games, and ebooks, probably with pictures, to make learning more enjoyable and effective.
2. Alignment with Educational Standards: This involves ensuring the curriculum complies with the Common European Framework of Reference for Languages (CEFR) and the UAE Ministry of Education.
3. Accessibility: This involves making the app available on major smart platforms like Android and iOS for widespread accessibility.
4. Self-Learning Capabilities: This is about allowing children to learn and play without adult supervision, furthering their independence in learning.

Challenges

Arabee faced myriads of challenges on its path to growth. They include;

1. Content Development: Crafting engaging and culturally relevant content aligned with educational standards was challenging as it required expert input from educators and linguists.
2. Technological Integration: Developing an app that is both user-friendly for children and capable of delivering all-encompassing

language education requires sophisticated technological solutions.

3. Market Penetration: Introducing a new EdTech solution in a market with established traditional language learning methods necessitated marketing and collaborations.

Outcomes

Arabee successfully solved the problems above and achieved remarkable outcomes;

1. Enhanced Learning Experience: Arabee's interactive platform has significantly improved engagement among learners, making Arabic language learning more enjoyable.

2. Comprehensive Curriculum: The app aligns with international standards and the Ministry of Education's requirements, ensuring that learners receive quality education.

3. Positive Reception: The app has enjoyed a warm reception from parents, educators, and learners, who have given positive feedback on its efficiency.

CHAPTER 7

AgTech – Innovative Agriculture through Technology

T he Middle East has faced multiple challenges in agriculture due to water scarcity, population increase, and the region's arid climate. These problems have put a lot of pressure on food security and agricultural viability. However, the increasing advancement and adoption of agricultural technologies (AgTech) are providing revolutionary solutions. Let's broaden the horizon on the transformative impact of AgTech in the Middle East, showing achievements, expansions, and the sector's future outlook.

The Middle East's harsh climate and scarce water supply have restricted agricultural productivity. Over 85% of food being imported in some Gulf Cooperation Council (GCC) countries is clearly at risk of global supply chain disruptions. Through the comprehension of such challenges, successive MEA governments have increasingly prioritized the continuous development and adoption of Agtech to advance food security, reduce dependency on imports, and boost agricultural practices.

Objectives

The Middle East's primary objective in integrating AgTech innovations is threefold:

1. Food Security Enhancement: reducing import dependency and boosting agricultural production locally.

2. Promoting Viable Agriculture: implementing practices that use resources more efficiently, especially water.

3. Agricultural Productivity Upgrade: there is a major increase in crop yield and quality through the implementation of crop technologies.

Implementations

1. Vertical Farming and Hydroponics: Several countries, like Saudi Arabia, are massively investing in vertical farming and hydroponics. These technologies allow year-round crop production with minimal water use. AeroFarms AgX in Abu Dhabi, the world's largest R&D indoor vertical farm, uses this method.

2. Controlled Environment Agriculture (CEA): the use of CEA technologies, including greenhouses facilitated with climate control systems, has made farmers otherwise hostile to agriculture while growing crops in certain conditions.

3. Precision Agriculture involves implementing IoT devices, drones, and AI that oversee crop health, improve water use, and predict agricultural outputs. This technology enables farmers to make decisions driven by data models, improving sustainability and productivity.

4. Genetic Engineering: drought-resistant and fast-growing crop varieties are highly developed here to combat the contrary effects of climate change and water scarcity.

Challenges

The adoption of AgTech in the Middle East has brought advancements, although there are several challenges across:

1. Exuberant Cost for Initial Investment: The capital required to implement advanced AgTech solutions can be extreme for small-scale farmers.

2. Technical Skills Gap: Training and education are required to equip local farmers and agricultural professionals with the essential skills to utilize AgTech innovations constructively.

3. Regulatory Hurdles: Navigating regulatory frameworks that may not be adapted to modern agricultural technologies can slow the adoption phase of AgTech integration.

Outcomes and Impact

The Middle East's adoption of AgTech has received some promising outcomes:

1. Increased Harvest Turnout: this is about multiplied productivity through suitability in agriculture and controlled environment farming.

2. Water Conservation: Water usage has been reduced significantly in a bid to address issues of water scarcity.

3. Sustainable Practices: This indicates a shift towards more viable and consistent agricultural methods that reduce environmental impact.

Future Outlook

The future of the Middle East in terms of AgTech is bright, especially in light of the continued investments and innovation pushing the sector forward. International collaborations are bringing global expertise to the region. Meanwhile, governments are actively promoting AgTech startups. Dubai's Food Tech Valley, an establishment of AgTech, stipulates a strong commitment to becoming leaders in enhanced agriculture.

AgTech innovations play a decisive role in transforming agriculture in the Middle East and beyond, granting solutions to longstanding challenges of water scarcity, climate change, and food security. The strategic integration of these technologies offers an assurance of a more self-sufficient and eco-friendly agricultural future. As the region keeps adopting AgTech solutions, the Middle East will invariably emerge as a top player in innovative farming practices and overcome its agricultural limitations.

Case Study - AEROFARMSAGX: Transforming Agriculture through AI and Plant Biology

AeroFarmsAgX: Pioneering the Future of Agriculture in the Middle East

As the world consistently struggles with the issues of climate change, water scarcity, and the shortage of sustainable food production, the MEA is fast looking for innovative solutions to secure its food future. Amongst these innovations, one stands out on a pedestal at the very top of agricultural technology: AeroFarmsAgX. The establishment is located in Abu Dhabi and it is also supported by the Abu Dhabi Investment Office. It is not just another farm but a research and development center that is on hand to transform farming as we know it.

The Creation of AeroFarmsAgX

AeroFarms is a top leader in the vertical farming technology industry. Incidentally, the company has long been at the forefront of developing solutions to some of the global agricultural problems. With the launch of AeroFarmsAgX, the company has taken a step forward by setting up the world's largest indoor vertical farm for research and development in Abu Dhabi. This shows an important partnership that is aimed at making use of the region-specific climatic challenges of the Middle East to bring up agricultural practices that can be applied globally.

Why Abu Dhabi?

Abu Dhabi is committed to furthering innovation in agriculture and ensuring food security. This is why it is the ideal location for AeroFarmsAgX. The region's arid climate and scarce water resources are a nagging challenge to traditional farming methods; thus, it is a suitable testing ground for AeroFarms' cutting-edge technologies. By situating its R&D center in Abu Dhabi, AeroFarms is contributing to the region's food security efforts and showing the potential of AgTech to thrive in severe environments.

Mission of AeroFarmsAgX

The mission of AeroFarmsAgX is multifaceted as it focuses on improving crop yield, conserving water, reducing the agricultural footprint, and, ultimately, redefining agricultural possibilities. The facility is dedicated to researching and bringing up new crop varieties that can withstand extreme climates. It also focuses on bringing up new farming techniques that use less water and no soil as well while browsing sustainable ways to feed the growing world population.

Innovations and Impact

AeroFarmsAgX is introducing several innovations in the field of vertical farming. These innovations include:

1. Aeroponics is a soil-less growing method in which roots are mixed with nutrients, water, and oxygen, resulting in faster growth rates and higher yields.
2. LED Lighting: There are customized LED lighting spectrums that can optimize plant growth and reduce energy consumption.
3. Data Analytics: This entails using big data and machine learning to analyze thousands of data points throughout the plant and animal growth process, thus inducing precision agriculture that can predict and enhance crop outcomes.

The impact of these innovations extends beyond the Middle East. As the environment around the world keeps on increasing, the work being done at AeroFarmsAgX becomes more impactful as it offers hope for sustainable agricultural practices that can be adopted globally, especially in regions facing similar problems.

Looking Ahead

AeroFarmsAgX is still keeping up with its actions in pushing beyond the boundaries of agriculture, and its work in Abu Dhabi will invariably have an immense impact on the future of food security, not in the Middle East alone but globally. The facility's research into water-efficient farming methods and crop resilience offers a blueprint for sustainable agriculture worldwide in arid and semi-arid regions.

AeroFarmsAgX symbolizes a forward leap in the search for sustainable, efficient, and resilient agricultural systems. By harnessing the power of AgTech, the facility addresses the immediate challenges of food production in the Middle East. Also, it contributes to the global conversation on how we feed our planet's growing population. The innovations developed at AeroFarmsAgX can coin the future of farming, and it is an exciting time for agriculture in the Middle East and worldwide.

AeroFarmsAgX is an excellent move in the global agricultural sector, especially in vertical farming and AgTech innovation. It is a perfect sample of how technology, sustainability, and agriculture can come together to address global problems like food security, water scarcity, and the need for sustainable farming practices. This model is not just about growing crops but also about reimagining the future of food production.

Business Model Overview

AeroFarmsAgX's model is based on the Controlled Environment Agriculture (CEA) in a vertical farming setup. This method makes for cultivating crops indoors, using stacked layers to maximize space. The major components of AeroFarmsAgX's business model are:

1. Technology-driven Cultivation: This involves the use of advanced aeroponics and hydroponics systems, LED lighting specifically tailored to plant needs, and climate control technologies to create the optimal growing environment for various crops, obviating the need for soil.

2. Data Analytics and AI: This is all about using cutting-edge data analytics and AI to monitor plant growth, health, and nutrient levels in real time to ensure yield and reduce waste.

3. Sustainability Focus: This entails stressing sustainability through reduced water usage compared to traditional farming, less use of pesticides, the ability to grow crops all year round, and reducing the carbon footprint associated with food transportation.

4. Scalability: this involves the design of farms near urban areas that can be scaled vertically and horizontally and also used to reduce distribution chains and make fresh produce accessible to consumers.

Revenue Streams

AeroFarmsAgX's model has a myriad of income streams, which include:

1. Direct Produce Sales: This involves selling premium-quality, pesticide-free leafy greens, and other crops directly to retailers, restaurants, and consumers.

2. Farming-as-a-Service (FaaS): This involves offering technology and expertise to other businesses looking to enter the vertical farming space, including setup, management, and ongoing support services.

3. Technology Licensing: This entails licensing its proprietary technology and systems to other vertical farms around the

world, creating a network of technology-driven sustainable farms.

4. Research and Development: This involves major collaborations with universities, government agencies, and private entities on agricultural research projects. It is funded through grants, partnerships, and direct contracts.

Challenges and Solutions

1. High Initial Investment: The upfront cost of establishing a vertical farm is vital, as without available finances, there'd be no farm. AeroFarmsAgX mitigates this through partnerships and investment rounds focused on long-term sustainability and profitability.

2. Energy Consumption: while LED lighting is more energy-efficient than traditional options, the energy demand remains high. This makes AeroFarmsAgX invest in energy-efficient technologies and explore renewable energy sources to power its farms.

3. Market Adoption: It is essential to educate consumers and retailers about the benefits of vertically farmed produce. AeroFarmsAgX focuses on marketing its products' superior quality, taste, and sustainability to induce demand.

AeroFarmsAgX's business model aptly shows the potential of AgTech to transform food production completely. Thus, by prioritizing sustainability, leveraging advanced technologies, and exploring new revenue streams, AeroFarmsAgX is not just growing crops but also nurturing the future of agriculture. As the world struggles with the challenges of feeding an ever-increasing population on a warming planet, models like AeroFarmsAgX offer a blueprint for sustainable, resilient, and efficient farming practices that could shape the future of food.

Case Study - PURE HARVEST FARMS: ReImagining Farming and Revolutionizing Food Production

Pure Harvest Farms – Innovation in Agriculture

The Pure Harvest Farm start-up has gained recognition for its innovative approach to agricultural improvement in severe environments. Here is a structured business case analysis based on widely recognized factors that contribute to the success of agricultural ventures like Pure Harvest.

Business Model and Value Proposition

Pure Harvest Farm's model is built around high-tech, controlled-environment agriculture (CEA), so it can produce fresh, high-quality produce all year round, irrespective of external climate conditions. This model is especially useful in regions with arid climates, where traditional farming is difficult or downright impossible. Their value proposition includes providing locally grown, pesticide-free, and nutrient-rich produce, thus reducing the need for imports, which can be expensive and less fresh.

Innovative Technology

Pure Harvest's success can be due to the adoption of advanced technologies, including hydroponics, aeroponics, or aquaponics, combined with precision climate control in greenhouses. This technology is effective for the useful growing conditions for various crops, leading to higher yield, faster growth, and low use of water and resources. The technology also supports sustainable farming practices by reducing the carbon footprint associated with traditional agriculture and food transport.

Market Demand

There is a rising global demand for fresh, locally sourced, and sustainable produce. Pure Harvest's success is partly due to its ability to meet this demand in markets previously reliant on imported goods. By providing a local alternative, Pure Harvest caters to individuals' desire for fresher produce and also appeals to environmentally conscious consumers looking to reduce their carbon output.

Sustainability Practices

Sustainability is the most important part of Pure Harvest's operations. Their use of CEA technology reduces water usage, essential in water-scarce regions. In addition, by obliterating the use of pesticides and reducing transport miles, they lower the negative environmental impact of their produce. These practices help in environmental sustainability and attract consumers and stakeholders to increasingly prioritize sustainability.

Partnerships

The company's ability to forge vital partnerships has been pretty sacrosanct to its success. Partnerships with technology providers, local governments, and investors have enabled Pure Harvest to upscale, access necessary capital, and navigate legal regulatory setbacks. The collaborative efforts with research institutions have also facilitated continuous innovation in farming practices and technology.

Pure Harvest Farm's success is due to its novel business model, which combines advanced agricultural technologies with sustainability and market demand for fresh, local produce. Using these factors, Pure Harvest has established itself as a high-tech, sustainable agriculture leader. As the global population continues to grow and climate change presents more significant challenges to agriculture, companies like Pure Harvest are needed to keep thriving around the globe by offering sustainable and innovative solutions to food production.

CHAPTER 8

Sustainability and Environment

I n the Middle East, the sustainability space is quite diverse owing to the region's geographic, climatic, and socio-economic activities. There are myriad challenges with the environment in the Middle East. The Arab governments have been working on numerous initiatives and innovative methods to drive sustainability. Some of the most significant challenges are water scarcity, desertification, greenhouse gas emissions, a dearth of biodiversity, and extreme weather changes. There is a growing momentum toward sustainability, with various creative innovations and strategies implemented across the region to address these adversities and promote environmental conservation and sustainable development. The aspects of this overview are as follows:

Challenges in the MEA Environment

1. Scarcity of Water: The Middle East is very water-stressed. Several countries in the area currently face severe water shortages, which can be caused by extraction, pollution, and inefficient usage. Climate change worsens the situation by altering precipitation patterns and increasing drought frequency.

2. Desertification: This is the process of gradual land degradation due to a combination of natural processes and human activities. The ecosystem of the MEA region is very fragile as encroachment threatens the remaining fertile land. Factors like increase in population, food consumption systems, and land degradation are leading to rampant desertification.

3. Greenhouse Gas Emissions: The region's economy depends greatly on oil and gas, leading to high per capita greenhouse gas emissions. The energy sector contributes the highest quota to the region's carbon.

4. Dearth of Biodiversity: Urbanization, pollution, and climate change are currently threatening the Middle East's biodiversity, which includes coral reefs in the Red Sea, mountains, and deserts.

5. Extreme Weather Changes: In April 2024, the Middle East (UAE and parts of Saudi Arabia mostly) was hit by heavy rainfall, which led to flood-ravaged neighborhoods, clogged highways, and submerged vehicles. One of the world's busiest airports, DXB i.e Dubai International Airport, came to a standstill, and several flights were canceled or rescheduled. A

team of global researchers presented a report attributing the heaviest rainfall in 75 years to Climate Change.

Initiatives to Build Sustainability

1. Renewable Energy: Solar and wind power currently make up the majority of renewable energy projects in Middle Eastern countries, which has helped to reduce dependence on fossil fuels and carbon emissions. Countries like the UAE and Saudi Arabia have set up programs to help raise the share of renewables in their energy mix.

2. Water Conservation: To solve water scarcity issues, numerous efforts are being carried out to improve water conservation and management by adopting modern irrigation techniques, desalination technologies, and wastewater recycling.

3. Sustainable Cities/Green Buildings: The MEA region is currently developing eco-friendly buildings to maximize environmental and economic impact. This is achievable via energy efficiency, sustainable materials, and waste reduction. Some examples are Masdar City in the UAE and Saudi Arabia's planned smart city, NEOM.

4. Marine and Terrestrial Conservation: Several conservation programs are currently being implemented to protect and restore endangered habitats like coral reefs, mangroves, and deserts. Measures being taken to conserve natural space include the establishment of protected areas, the use of sustainable fishing practices, and the rejuvenation of degraded lands.

5. Climate Action: Middle Eastern countries' implementation of action plans is part of their commitments under the Paris

Agreement. These plans include measures to modify and reduce climate change's effects.

Economic Diversification

Economic diversification is a critical component of the sustainability agenda in the Middle East. Numerous countries actively seek to narrow their reliance on oil and gas revenues, thereby handling renewable energy, tourism, technology, and sustainable agriculture as investment proteges. This shift is seen as essential for long-term economic stability and environmental sustainability.

The MEA's approach to sustainability and environmental challenges is changing quickly. While the region faces significant hurdles, the increasing commitment of governments, businesses, and civil societies to environmental conservation and sustainability is a positive picture.

The Middle East aims to address environmental issues through innovation, policy reform, and international cooperation, thus contributing to global sustainability.

Notable Names

1. **Solavio Labs**: Solavio Labs is a technology company focused on advancing solar power through innovative engineering solutions, with a particular emphasis on developing products like their autonomous cleaning bot to enhance the performance and efficiency of solar power plants. Engineers founded it to develop innovative technologies and engineering solutions to improve the performance of solar power plants. One of their solutions is an autonomous solar panel cleaning bot with a modular design, making it compatible with various structures, mounting areas, and climatic conditions. The company is

continuously growing and developing new designs, patenting innovative solutions, and engineering new products.

2. **QANATU**—Qanatu is a technology platform based in Dubai that enables bottled water companies to streamline and enhance their home delivery services. It leverages advanced technology to improve efficiency, competitiveness, and sustainability while providing an excellent customer experience and valuable analytics on consumer behavior and sales data.

CHAPTER 9

Renewable Energy Landscape

The renewable energy landscape in the Middle East is undergoing massive transformation, purely driven by strategic economic diversification attempts, abundant solar resources, and a growing realization of the need to hold back climate change. Once recognized as reliant on fossil fuels, the Middle East has now repositioned itself as a universal leader in renewable energy development, notably in solar and wind power. The following overviews result from the region's renewable energy landscape.

Solar Energy Dominance

1. Prolific Solar Resources: the world's high-rise solar irradiance levels are bestowed in the Middle East regions. This measure gives an exceptional option for the region producing solar energy. The United Arab Emirates (UAE), Saudi Arabia, and Jordan have a worthwhile number of solar energy projects. This

is why UAE's Mohammed bin Rashid Al Maktoum Solar Park and Saudi Arabia propose such remarkable innovations of Vision 2030.

2. Solar Technologies Revolution: the region focuses on expanding its solar energy capacity and constructing solar technologies. Traditional Photovoltaics (PV) systems are established with Concentrated Solar Power (CSP) projects, which can store energy for use when it is cloudy.

Potential Wind Energy

1. Emerging Wind Projects: While solar energy is the most prominent, wind energy is also gaining interest in the Middle East. Jordan and Saudi Arabia are exploring wind energy's potential, with several projects already operational or in the pipeline. Wind farms are generally satisfactory for the region's hilly terrains and coastal areas.

Energy Transformation Economic Variety

1. Shifting away from Oil: Many Middle Eastern countries seek to diversify their economies away from dependence on oil. Renewable energy development is the central system that will engage in diversification, reducing domestic fossil fuel consumption and freeing oil and gas for export.

2. Visionary Projects: Projects like NEOM in Saudi Arabia and Masdar City in the UAE exemplify the region's commitment to integrating renewable energy and reliable technologies into new urban areas and economic zones.

Government Policies and Investments

1. Supportive Policies: Governments across the Middle East are implementing policies to support government renewable energy investments, including feed-in tariffs and auctions for solar and wind projects.

2. International Collaboration: International renewable energy organizations fund the projects and offer expertise to the regions seeking partnerships.

Barriers and Opportunities

1. Infrastructure and Grid Integration: As renewable energy capacity extends, barriers related to grid integration and storage become more pronounced. Grid infrastructure investment and energy storage solutions are required to maximize renewable energy potential.

2. Economic and Social Benefits: Economic opportunities such as technological innovation and jobs are offered through the renewable energy sector. It also provides social benefits, including reducing air pollution and combating climate change through global efforts.

Future Outlook

The Middle East's renewable energy space is ready for rapid growth, with solar energy paving the way and wind energy quickly adapting. This development supports both economic diversification sustainability objectives and environmental goals. As technologies push forward and investments continue, the Middle East could become a global benchmark, achieving a traditionally fossil-fuel-dominated economy by integrating renewable energy. This shift towards renewables is critical to the region's broader efforts to ensure long-term energy security and economic and environmental sustainability.

Notable Names

1. Desolcenator—The world's first circular solar desalination system enhances clean water production from complex sources by tapping into the sun and abundant sea sources.
2. Enerwhere - Data-driven solar utility company providing clean, reliable energy with modular solar-hybrid mini-grids.

The Secret Ingredient

Embracing Your Journey

The destination through enlightenment has reached its last bus stop.

Congratulations!

You have embarked on a comprehensive journey through the vibrant landscape of technology start-ups reforming the Middle East and the broader Arab world. Drawing back on regional challenges to be addressed by innovative solutions to groundbreaking platforms capturing global recognition, the stories you have encountered indicate the power of vision and determination.

You have also witnessed diverse features, business models, and strategies that show a dynamic region flourishing in the spirit of entrepreneurship. Each tale is a unique blueprint that describes how to launch a start-up and how to navigate the market's complexities, innovate continuously, and reconstruct to ever-changing consumer needs and technological advancements.

The principal message here is powerful and distinctive: boldness is the cornerstone of entrepreneurship. It is about the audacity to dream and the courage to pursue those dreams relentlessly. It is about making tough decisions when the path before you is unpredictable and vague, but at least staying true to your vision.

Remember, the journey of entrepreneurship is unpredictable. It is filled with highs and lows, successes and setbacks. Yet, it is critical to know that money, capital, and human resources, although essential, are secondary investments brought due to your sheer will and perseverance. These resources will eventually align with your vision if you maintain your resolve and keep pushing forward.

Take calculated risks, but don't deny the possibility of failure. These risks can lead to growth and learning. Draft detailed models for your venture, and remember to be flexible enough to pivot when necessary. The ability to adapt is a hallmark of successful entrepreneurs.

Most importantly, stick to your plan with steadfast commitment. This is the secret ingredient. Believe that your aspirations will inspire others to believe in your dreams, driving the right forces, networks, and investors to your establishment. Persistence and dedication are undoubtedly the path to success.

As you push forward, let the stories of these pioneering start-ups fuel your ambition. The strategies and intuition of the most successful entrepreneurs of the Middle East and beyond must indeed encompass your mind. It's your turn to make waves, disrupt industries, and generate a considerable impact.

The world of entrepreneurship is ripe with possibilities. Dare to dream big, stay bold in your pursuits, and always remember that the most extraordinary achievements often come from taking the biggest risks. This journey has just started, and the future is yours to shape. Go out there and transform your vision into reality.

Those disheartening moments are not the signal to abandon your dreams but rather broaden your growth and learning process. The statement, "You have to see failure as the beginning and the middle, but never entertain it as an end," encompasses a robust and resilient mindset for personal development and long-term success.

All the best!

Notes

Part 1

Chapter 1

https://www.itu.int/en/ITU-T/Workshops-and-Seminars/sg05rg/sdtd/20210929/Pages/default.aspx

https://www.forbes.com/sites/forbestechcouncil/2023/04/17/digital-transformation-in-the-middle-east-challenges-and-opportunities/

Chapter 2

https://wam.ae/en/article/13rktma-digital-transformation-pillar-arab-strategies

https://www2.deloitte.com/xe/en/pages/technology-media-and-telecommunications/articles/dtme_tmt_national-transformation-in-the-middleeast-a-digital-journey.html

https://www.weforum.org/agenda/2023/07/is-the-middle-east-showing-how-the-digital-revolution-will-be-people-powered/

https://wired.me/technology/middle-east-technology-trends-from-morocco-to-oman/

https://www3.weforum.org/docs/WEF_Digital_Arab_World_White_Paper_2018.pdf

https://www.jstor.org/stable/4329507

https://digitalcommons.macalester.edu/cgi/viewcontent.cgi?referer=&httpsredir=1&article=1415&context=macintl

Chapter 3

https://en.incarabia.com/top-rising-startups-in-the-arab-world-603622.htmlhttps://www.forbes.com/sites/afshinmolavi/2023/02/15/middle-east-tech-startups-are-a-hot-emerging-market/

https://www.linkedin.com/pulse/booming-startup-ecosystem-mena-raha-beach-ventures

https://www.dw.com/en/saudi-arabias-thriving-startup-scene-driven-by-women/a-68569509

Chapter 4

https://www.thehartford.com/business-insurance/strategy/how-to-start-a-business/startup

https://www.vationventures.com/blog/what-makes-a-successful-startup

https://www.forbes.com/sites/allbusiness/2018/07/15/35-step-guide-entrepreneurs-starting-a-business/

https://hbr.org/2019/07/building-a-startup-that-will-last

Part II

Chapter1

https://www.arabnews.com/node/2430886/business-economy

https://www.arabnews.pk/node/2429671/business-economy

https://www.fintechfutures.com/2023/12/bnpl-fintech-tabby-lands-700m-debt-facility-from-jp-morgan/

https://fintechnews.ae/19476/saudiarabia/tabby-secures-usd-700-million-in-debt-financing-and-extends-series-d-round-to-usd-250-million/

https://www.wamda.com/2024/03/fawry-dahab-partners-payme-facilitate-egyptian-expats-transactions

https://fintech.global/2023/12/15/fawry-and-hulul-unite-to-empower-smes-with-ai-enhanced-e-payment-solutions/

https://yourstory.com/2024/03/fawry-and-payme-digital-collaborate-to-serve-egypt

https://www.cnn.com/videos/tv/2024/01/10/mpa-profit-point-fawry.cnn

https://ibsintelligence.com/ibsi-news/optasia-to-provide-micro-lending-solution-to-bede-bahrain/

https://techlabari.com/digital-lending-company-optasia-obtains-fintech-license-from-bank-of-ghana/

https://fintech.global/2023/11/16/paytabs-partners-with-tabby-to-power-ecommerce-growth-across-saudi-arabia/

https://www.zawya.com/en/press-release/companies-news/paytabs-group-inks-partnership-with-fintech-galaxy-to-elevate-gccs-fintech-space-with-payment-orchestration-and-open-banking-solutions-gjtrkx9w

https://www.arabnews.com/node/2427826/business-economy

https://www.fintechfutures.com/2023/12/saudi-bnpl-fintech-tamara-lands-340m-series-c-at-unicorn-valuation/

Chapter 2

https://www.arabnews.pk/node/2441066/business-economy

https://www.pymnts.com/news/b2b-payments/2022/b2b-ecommerce-platform-sary-expands-pakistan-stake-jugnu/

https://techcrunch.com/2023/12/22/maxab-and-wasoko-in-merger-talks/

https://www.arabnews.com/node/2434371/business-economy

https://www.arabnews.com/node/2146816/business-economy

https://tracxn.com/d/explore/e-commerce-logistics-startups-in-middle-east/__gtQjVKWJaAY3ai-RVdQWxcvjYoMtP-mLKBmamRvSonI/companies

https://www.arabianbusiness.com/money/corporate/funding/saudi-based-e-commerce-retailo-raises-36-million-in-series-a-round

https://www.forbesmiddleeast.com/lists/top-50-most-funded-startups/

Chapter 3

https://techcrunch.com/2022/06/28/egyptian-healthtech-startup-vezeeta-cuts-10-of-500-person-staff/

https://www.connectingafrica.com/author.asp?section_id=781&doc_id=767008

https://www.mobihealthnews.com/news/uae-digital-health-startup-altbbi-scores-44m

https://www.thenationalnews.com/business/start-ups/2022/05/02/generation-start-up-how-altibbi-is-bringing-health-care-to-the-home/

https://www.wamda.com/2023/03/saudi-arabia-aumet-raises-7-million-pre-series-round

https://waya.media/healthtech-onex-raised-us1-2m-pre-seed-funding-round/

https://www.business-standard.com/companies/start-ups/health-tech-start-up-neodocs-aims-to-do-100-million-tests-in-3-years-124022500375_1.html

https://www.financialexpress.com/business/healthcare-ai-powered-diagnosis-apps-signify-transformative-shift-in-healthcare-delivery-3402810/

https://www.thenationalnews.com/future/technology/2024/05/27/how-clinicy-aims-to-efficiently-connect-medical-service-providers-with-patients/

https://www.arabnews.com/node/2457686/business-economy

Chapter 4

https://www.edgemiddleeast.com/industry/startup-edge-from-offline-to-online-huspys-mission-to-digitise-real-estate-in-mena

https://www.zawya.com/en/press-release/companies-news/proptech-huspy-announces-record-aed-1bln-in-mortgages-processed-in-november-2023-aht7geak

https://tracxn.com/d/explore/real-estate-tech-startups-in-united-arab-emirates/__pZYZqJy8TZJ-nI90oG3yjbHvvK3f-r9aWvsSHqJIIJo/companies

https://www.zawya.com/en/press-release/companies-news/bayuts-q1-2024-data-shows-stable-apartment-prices-amidst-continued-property-surge-mz9b7bhz

Chapter 5

https://www.arabnews.com/node/2053771/business-economy

https://www.dailynewsegypt.com/2024/01/04/swvl-rehires-some-of-its-employees-expands-its-services-in-saudi-arabia-in-2024/

https://www.logisticsmiddleeast.com/news/swvl-to-carefully-expand-in-saudi-arabia-in-2024

Chapter 6

https://www.arabnews.com/node/2402041/business-economy

https://www.wamda.com/2023/11/edtech-noon-closes-41-million-series-b

https://economymiddleeast.com/news/edtech-startups-in-saudi-arabia/

https://gulfnews.com/uae/education/app-launched-in-uae-to-make-learning-arabic-fun-and-easy-1.75021368

Chapter 7

https://agfundernews.com/aerofarms-emerges-from-ch-11-as-its-virginia-based-indoor-farm-nears-profitability

https://amore.ng/news/32/vertical-farming-produce-market-emerging-trends-aerofarms-agricool-badia-farms/

https://www.just-food.com/news/aerofarms-emerges-from-bankruptcy-process-with-new-ceo-molly-montgomery-appointed/

https://www.wamda.com/2023/12/pure-harvest-acquires-red-sea-production-facility-saudi-arabia

https://ess.honeywell.com/us/en/about-ess/newsroom/press-release/2021/04/universal-studios-beijing-adopts-honeywells-environmentally-preferable-solstice-n40-refrigerant

Chapter 8

https://www.arabnews.com/node/2494351/saudi-arabia

https://www.globalcompliancenews.com/2024/03/18/https-insightplus-bakermckenzie-com-bm-environment-climate-change_1-united-arab-emirates-regulations-on-single-use-products-environmental-sustainability-in-dubai_02272024/

https://www.undp.org/arab-states/stories/rising-challenge-climate-action-arab-region

Chapter 9

https://solarquarter.com/2024/01/22/arab-nations-shine-in-renewable-energy-a-unesco-science-report/

https://www.undp.org/arab-states/stories/unleash-power-renewable-energy-arab-region

https://www.arabnews.com/node/2508326

https://energy.ec.europa.eu/news/commissioner-simson-international-renewable-energy-agency-assembly-foster-global-cooperation-2024-04-16_en

ABOUT THE AUTHOR

Dr. Uzma Farheen is a versatile individual and a true inspiration with a wide range of interests encompassing medicine, business strategy, public health, and creative writing.

Dr. Uzma, a medical doctor by training, has uniquely enriched her career with a dynamic blend of healthcare consulting experience and business strategy. She earned a postgraduate degree from the prestigious Indian School of Business and is pursuing a Master's in Public Health at Harvard University.

Dr. Uzma's personal journey is a powerful testament to resilience and courage. Overcoming profound personal challenges, she restarted her life with virtually nothing, completing her education on scholarships and rising above obstacles with grace and determination. Her story is a beacon of hope, embodying the indomitable power of the human spirit to overcome adversity and inspiring others to persevere in their own struggles.

Dr. Uzma's life extends beyond her professional and academic pursuits. She is a devout wife and a loving pet parent, finding joy and solace in her faith and the companionship of her furry friends. Her love for baking and exploring new cultures through travel speaks to her open heart and boundless curiosity, inviting others to share in her diverse experiences.

Dr. Uzma is a firm believer in love, kindness, and empathy. With her unique perspectives and a heart brimming with warmth and understanding, she is passionately dedicated to captivating and inspiring readers with her storytelling. Her narratives are meticulously crafted to uplift and encourage, offering a powerful message of hope and a heartfelt celebration of the beauty in resilience.